Wetland Restoration

Springer Environmental Science and Engineering

For further volumes:
http://www.springer.com/series/10177

Shuqing An • Limin Wang

Wetland Restoration

Shanghai Dalian Lake Project

Science Press
Beijing

Springer

Shuqing An
School of Life Science
Nanjing University
Nanjing, China

Limin Wang
Society of Entrepreneurs and Ecology
World Wide Fund for Nature
(WWF) China
Shanghai, China

ISSN 2194-3214 ISSN 2194-3222 (electronic)
ISBN 978-3-662-51088-9 ISBN 978-3-642-54230-5 (eBook)
DOI 10.1007/978-3-642-54230-5
Springer Heidelberg New York Dordrecht London

Jointly published with Science Press, Beijing
ISBN: 978-7-03-040057-4 Science Press, Beijing

Softcover reprint of the hardcover 1st edition 2014

Printed on acid-free paper

Springer is part of Springer Science+Business Media (www.springer.com)

Foreword

I am glad to know the book titled *Wetland Restoration—Shanghai Dalian Lake Project* would be published very soon. Dalian Lake restoration project is a water source rehabilitation demonstration project launched by our cooperative partner, World Wide Fund for Nature (WWF), under the support of the HSBC Climate Partnership program; in view of the concerns about the drinking water safety of the megalopolis of Shanghai, the project was launched for trying to solve the agricultural nonpoint source of pollution and domestic community pollution in the water source area.

As the sponsor of the project, I was lucky to make a personal survey on the project in the field of Dalian Lake, and I am impressed by it: after rehabilitation, the wetland has not only acquired a beautiful environment but also a clean water source; the local communities participated in the water source area protection by launching organic agriculture, which realized the win-win situation of water source protection and economic benefit. Such a "1 + 1 partner in water source area" mode, namely the mode that attracts and absorbs the local communities, government, experts, NGOs, and enterprises, is very worthy of popularization for wetland rehabilitation, popularization for organic agriculture, and urban water source purification.

I am also glad to know that the project had won multiple honors after it was implemented successfully, such as "The first Occasion of Shanghai Award for Protection of Mother River," "Excellent Proposal Award of Shanghai Municipal CPPCC," and "Best Project of the first Occasion of Foundation for Guangcai Program." I think this is the affirmation of the government and all circles in the society for the achievements of the Dalian Lake project.

In June 2012, HSBC launched a new round of the 5-year-term HSBC Water Resource Program with WWF and other cooperative partners, aiming to jointly boost the protection, preservation, and sustainable utilization of the water resources in the middle and lower reaches of the Yangtze River. We believe that the future of the Yangtze River basin is very important to the economic growth of China. By being established in the enterprise's sustainable development, HSBC is willing to boost the harmonious and healthy development of economy, community, and enterprise with our cooperative partner WWF through species protection, wetland restoration, enterprise participation in water management, environment-friendly fishery industry, and integrated watershed management and then finally help out with the green transformation of the Yangtze River basin.

To invest in water is to invest in future. I sincerely express my best wishes for the publishing of the book.

HSBC Bank (China) Company Limited
Shanghai, China
October 2013

Bijuan Huang

Preface

After the Jiangsu Jiangyan Qin Lake "Wetland Forum" conference was convened in April 2008, Dr. Wang Limin, the deputy director of the World Wide Fund for Nature (WWF) China Programme Implementation and the director of World Wide Fund for Nature (WWF) Shanghai Programme Office, came to my laboratory in Nanjing University for visiting and communication. Dr. Wang was impressed on the achievements of the wetland restoration engineering completed by the ecology department of Nanjing University; after that, Dr. Wang invited me and my wetland restoration engineering team to go to the Shanghai Dianshan Lake for conducting a similar scientific research work so as to finally make a contribution to the protection of the water source area of Huangpu River in Shanghai City.

With the support of the HSBC Climate Partnership China Programme, the wetland restoration engineering team of Nanjing University conducted a systematic and detailed survey with regard to the ecological environment, social economy, and natural geography of the Shanghai Dalian Lake and the area where the program will be launched in July 2008 and its surrounding area and prepared the *Feasibility Study Report on Science & Technology Demonstration Programme for Shanghai Dalian Lake Wetland Restoration*. On this basis, with the full support of the Shanghai Lake Construction and Development Co., Ltd. (formerly Shanghai Dianshan Lake Development Co., Ltd.), and the Shanghai Administration for Afforestation and City Appearance, the wetland restoration engineering team of Nanjing University completed the *Protection Planning for Wetland Restoration and Water Source Area in Shanghai Dalian Lake* (2,000 mu[1]), *Construction Scheme Design for Shanghai Dalian Lake Wetland Restoration* (625 mu), and *Shanghai Dalian Lake Wetland Restoration Engineering Based on Community Participation* (150 mu), among which the first two were sponsored by the Shanghai Municipal Development & Reform Commission and were almost completed and the last one was sponsored by WWF, which is the first "tough" subject in the field of engineering in the sponsorship history of WWF.

[1] 1 mu = 667 m^2.

With the active collaboration and assistance of the relevant departments of the Shanghai Municipality and the Qingpu District Government and its related organizations, the 150 mu wetland restoration engineering has achieved satisfactory results. The acceptance and summary conference was convened in September 2010. At that time, the Water Source Area Forum was held in the WWF exhibition hall of Shanghai Expo, which was highly acclaimed by the leaders and experts of State Forestry Administration, State Water Resources Administration, Ministry of Environmental Protection, and relevant departments of the Shanghai Municipal Government. The senior experts of WWF International, WWF Britain, WWF China, HSBC, and the external senior assessment experts were invited to visit the construction field and conduct field monitoring during the construction period; after analyzing the reports, all experts highly commented on the project and recognized it as a leading example in the world for the ecological rehabilitation of water source areas in megalopolises, having far-reaching significance.

This book is compiled on the basis of the above projects. It provides a detailed description for every stage of the wetland restoration project, including the wetland background investigation, overall planning of wetland rehabilitation, detailed design, project construction, tracking, monitoring, and assessment of engineering results. Upon the living example of the Shanghai Dalian Lake Wetland restoration project, the book furnishes firsthand data for the wetland restoration project construction and provides a complete and integral description for the overall implementation process of the wetland restoration project. This book can be used as a reference for similar wetland restoration projects in future.

Nanjing, China
December 2013

Shuqing An

Acknowledgments

The integration of the wisdom of the Shanghai Huangpu River water source area protection team and the selfless dedication of all partners has made possible the publishing of the book *Wetland Restoration*: *Shanghai Dalian Lake Project*. I highly and sincerely appreciate and would like to thank all institutions and individuals that supported the program on behalf of all team members.

Shanghai Administration for Afforestation and City Appearance: Youming Cai, Yimin Xie, Zhitong Zhang, and Cheng Xue
Shanghai Lake Construction and Development Co., Ltd.: Jinhua Shen, Ruiyun Zhang, Shuigen Chen, Xiaowei Zhang, and Chunhua Chen
Shanghai Wild Animal and Plant Protection and Management Station: Enle Pei and Shuzhong Xia
Shanghai Fishery Machinery and Instrument Research Institute, Chinese Academy of Fishery Sciences: Xingguo Liu
School of Life Sciences, East China Normal University: Tianhou Wang
Shanghai Garden Science Research Institute: Xinhong Cui
Shanghzai Academy of Environmental Sciences: Xianggen Shen and Yuchi Huang
Shanghai Oasis Ecological Protection Exchange Center: Bing Li and Siwei Chen

Shanghai Huangpu River Water Source Area Protection Team
World Wide Fund for Nature
Nanjing University
Fudan University
Shanghai Jiao Tong University

Acknowledgements

Contents

Part I Project Implementation Background

1 Locational Features of Dalian Lake 3
1.1 Geographical Location 3
1.2 Natural Conditions 4
1.3 Social Economic Conditions 8
Bibliography 12

2 Regional Evolution of Dalian Lake 13
2.1 Regional Wetland Type 13
2.2 Regional Wetland Biology 15
2.2.1 Plankton 15
2.2.2 Aquatic Vascular Bundle Plants 17
2.2.3 Zoobenthos 18
2.2.4 Fishes 19
2.2.5 Birds 19
2.2.6 Animals 19
2.2.7 Endangered Rare Species 19
2.2.8 Invasive Species 20
2.3 Regional Wetland Evolution 20
2.3.1 Kinds of Phytoplankton and Evolution of Dominant Species 21
2.3.2 Kinds and Quantity Evolution of Zooplankton 21
2.3.3 Type, Quantity and Distribution Evolution of Aquatic Vascular Bundle Plants 21
2.3.4 Type, Quantity Distribution and Evolution of Bentonic Organism 24
2.3.5 Composition, Distribution Characteristics and Aquatic Product Types of Fishes in Dianshan Lake 25
Bibliography 26

3 Characteristics of Hydrologic Environment 27
3.1 Hydrologic Characteristics 27
3.2 Physical and Chemical Characteristics of Water Environment 29
3.3 Biological Characteristics of Water Environment 32
3.3.1 Density of Phytoplankton 32
3.3.2 Chlorophyll *a* Content of Phytoplankton in Dalian Lake 34
3.3.3 Density of Zooplankton 34
3.3.4 Density of Zoobenthos and Biomass in Dalian Lake 35

4 Analysis on Pollution Source 37
4.1 Internal Source Pollution 38
4.2 Non-point Source Pollution 40
4.2.1 Farm Chemical Fertilizer Pollution 40
4.2.2 Animal Husbandry Pollution 40
4.2.3 Fishery Breeding Pollution 41
4.3 Point Source Pollution 41
4.3.1 Pollution by Rural Life 41
4.3.2 Pollution by Industries 41
Bibliography 42

Part II Project Planning, Design and Construction

5 Wetland Restoration Strategy 45
5.1 Evaluation on Status of Dalian Lake Wetland 45
5.1.1 Analysis on Ecological Functions of Dalian Lake Wetland 45
5.1.2 Integrated Evaluation on Status Quo of Dalian Lake Wetland 48
5.2 Wetland Restoration Thinking 50
5.2.1 Effect After Dalian Lake Wetland Is Rehabilitated 50
5.2.2 Restoration Strategy of Dalian Lake Wetland 51
5.2.3 Management Measures of Wetland 53
5.3 Philosophy and Principle for Planning and Designing on Wetland Restoration Project 54
5.3.1 Design Philosophy of Wetland Restoration 54
5.3.2 Design Principle of Wetland Restoration 55
5.4 Overall Planning for Wetland Restoration Project 56
5.4.1 Project and Its Significance 56
5.4.2 Project Planning Objective and Overall Thinking 57
5.4.3 Planning Content and Scale of Project 57
5.4.4 Anticipated Effect of Project 58
Bibliography 61

6 Wetland Restoration Project Layout and Subarea Design ... 63
6.1 General Engineering Layout ... 63
6.1.1 Wetland Restoration and Reconstruction Area (Subarea A) ... 63
6.1.2 Forest Wetland Cultivation Area (Subarea B) ... 64
6.1.3 Shoal Wetland Restoration and Biodiversity Conservation Area (Subarea C) ... 65
6.2 Engineering Subarea Design ... 65
6.2.1 Design of Wetland Restoration and Reconstruction Area 65
6.2.2 Design of Forest Wetland Cultivation Area ... 72
6.2.3 Design of Shoal Wetland Restoration and Diversity Conservation Area ... 74
6.3 Water Conservation and Earthwork Design of the Project ... 76
6.3.1 The Work Conservation Design of the Project ... 76
6.3.2 Earthwork Project ... 82

7 Other Designs for Wetland Restoration Project ... 83
7.1 Construction Organization Design ... 83
7.1.1 Engineering Conditions ... 83
7.1.2 Drained Waters in Construction Period ... 85
7.1.3 Construction Methods ... 85
7.1.4 Construction Layout ... 86
7.1.5 Soil Balance ... 87
7.1.6 Construction Area ... 87
7.1.7 Construction Progress ... 87
7.1.8 The Main Machinery Equipment for the Construction ... 88
7.1.9 The Main Technical and Economic Indexes for the Construction ... 88
7.2 Environment Protection Design ... 88
7.2.1 Environment Protection Goals ... 89
7.2.2 Environment Protection Standard ... 90
7.2.3 Analysis and Appraisal for the Affection of Project Construction on Environment ... 90
7.2.4 Environment Protection Design ... 93
7.2.5 Environmental Management and Monitoring Plan ... 95
7.3 Water and Soil Conservation Design ... 96
7.3.1 Design Basis ... 96
7.3.2 Fundamental Principle ... 97
7.3.3 The Status Quo of Water and Soil Loss ... 97
7.3.4 Earthwork Balance ... 98
7.3.5 The Responsibility Scope and Subarea of Water and Soil Loss Prevention and Control ... 98
7.3.6 The Analysis of the Affection of Development and Construction Activities on Water and Soil Loss ... 99
7.3.7 The Goal of Water and Soil Loss Prevention and Control ... 99

7.3.8 Measures of Water and Soil Loss Prevention and Control ... 100
7.3.9 Water and Soil Loss Monitoring ... 102
7.4 Project Management and Maintenance ... 102
7.4.1 Management Institution and Personnel Composition ... 102
7.4.2 Management Scope and Contents ... 103
7.4.3 Water Quality Monitoring and Biological Investigation ... 104
7.4.4 Annual Operation and Management Cost ... 104
7.4.5 Project Operation and Management ... 105
7.4.6 Planting and Management Methods of Wetland Plants ... 106
7.4.7 Management Laws and Regulations ... 108

8 Construction and Management of the Project ... 109
8.1 Construction Overview of the Project ... 109
8.1.1 Basic State of the Project ... 109
8.1.2 Construction Characteristics of the Project ... 109
8.1.3 Overall Construction Layout ... 110
8.2 Construction Organization and On-site Management ... 110
8.2.1 Construction Organization Structure ... 110
8.2.2 Construction On-site Management ... 112
8.3 Construction Solution of the Project ... 112
8.3.1 Main Construction Procedure ... 112
8.3.2 Construction Preparation and Temporary Facilities ... 113
8.3.3 On-site Clean Up ... 114
8.3.4 Earthwork Digging and Refilling ... 114
8.3.5 Quick Percolation System Filling ... 115
8.3.6 Plant Planting ... 116
8.4 Construction Progress and Guarantee System of the Project ... 116
8.4.1 Construction Progress and the Goals of Construction Period ... 116
8.4.2 Quality Guarantee System and Guarantee Measures of the Project ... 116
8.4.3 Safety Guarantee System and Safety Guarantee Measures ... 118
8.4.4 Technical Measures of the Construction in Winter and Monsoon ... 119
8.4.5 On-site Civilized Construction Measures ... 120
8.4.6 Construction Management and Others ... 120
8.5 Supervision of the Project ... 120
8.5.1 Supervision Goals and Guidance ... 120
8.5.2 Supervision Institution ... 122
8.5.3 Task and Method of Investment Control ... 123

8.5.4 Tasks and Methods of Progress Control 124
8.5.5 The Tasks and Methods of Quality Control 125
8.5.6 The Supervision Method of Defect Responsibility Period 126

Part III Effect Monitoring and Appraisal

9 Wetland Water Quality Monitor and Appraisal 129
9.1 The Water Quality Monitoring Layout and Appraisal Methods of the Project 129
9.1.1 Monitoring Layout and Time 129
9.1.2 Water Quality Monitoring and Appraisal Methods 129
9.2 The Monitoring Results of the Water Quality of the Project 131
9.2.1 The Physical Property of the Water 131
9.2.2 The Chemical Property of Water 133
9.2.3 The Biological Nature of Water 139
9.3 Other Monitoring Results of Wetland 142
9.3.1 Monitoring Results from Commissioned Unit 142
9.3.2 Monitoring Results from Other Units 142
Bibliography 147

10 Benefit Appraisal of the Wetland 149
10.1 The Ecological and Environmental Benefit of the Project 149
10.1.1 Ecological Benefit 149
10.1.2 Environmental Benefit 150
10.2 Social and Economic Benefit of the Project 151
Bibliography 153

Appendices 155
Appendix 1 155
Appendix 2 159
Attachment 2.1. Environment Quality Standard for Surface Water of the People's Republic of China 159
Attachment 2.2. The Implementation Guidelines of the Protective Regulations of the Water Source of the Upper Reaches of Huangpu River in Shanghai 174
Appendix 3 183
Attached Figures 1. Overall Plan Maps of Shanghai Dalian Lake Wetland Restoration Project (2,000 mu) 183
Attached Figures 2. Design Drawings of First-Stage Wetland Restoration Project (625 mu) 195
Attached Figures 3. Effect Figures of Wetland Restoration Project (150 mu) 212

Postscript & Testimonials 215

Part I
Project Implementation Background

Chapter 1
Locational Features of Dalian Lake

1.1 Geographical Location

Dalian Lake wetland restoration zone is situated in the south of Lanlu Port, 3.5 km in the southwest of downstream zone of Dianshan Lake, Qingpu District, Shanghai City, with the total land of 14.6 km^2 and the core area of 4.6 km^2 (Fig. 1.1). There are nine natural villages of two towns, namely Jinze Town and Zhujiajiao Town in the wetland restoration zone, where the total population is 7,504 (2,650 households). Dalian Lake District is 58 km away from Shanghai People's Square and Hu-Qing-Ping Highroad run through the zone to offer a convenient and quick transport (Fig. 1.2). Water from Dianshan Lake, getting through Xietang (Lanlu Port), converges with Yuanxiejing and Damao Port in Songjiang River, all three of them constitute Huangpu River; the Lanlu port runs through the whole region of Dalian Lake Zone. According to *Regulations of Shanghai Municipality for Preservations on Upstream Water Sources of Huangpu River*, the zone is an important water source protection zone of Shanghai Municipality, and its local ecological conditions have important strategic significance for Shanghai's sustainable development.

The water source area of Dalian Lake is an integral part of Dianshan Lake water system; where the latter is an alternately inflow-outflow lake in Tai Lake Zone, with the water area of 62 km^2, average water depth 2.1 m and maximum water depth 3.6 m. Dianshan Lake receives mainly the water from Tai Lake; where the water from Tai Lake flows into Dianshan Lake via Jishui Port, Dazhushe and other harbours from Northwest to Southeast, and then discharges into Huangpu River through Lanlu Port, Dianpu River and other rivers; generally it dwells about 29 days; the Tai Lake water is about 17 % of total water yield of Huangpu River and is one of main water sources of Shanghai Municipality (Fig. 1.3). The water from Dianshan Lake flows gently, with a flow speed of 0.03 m/s approximately; Jishui Port and Dazhushe are the main water intakes of Dianshan Lake, with the water yield of 35 % and 33 % of the total water inflows respectively; Lanlu Port is the main water outlet of Dianshan Lake, and its water yield is 71 % of the total water output. Dianshan Lake is a tidal lake, its water level and water yield are not only

S. An and L. Wang, *Wetland Restoration: Shanghai Dalian Lake Project*,
Springer Environmental Science and Engineering, DOI 10.1007/978-3-642-54230-5_1,

Fig. 1.1 Location of Dalian Lake

affected by upstream water, but also by the tidal level of Huangpu River. Dianshan Lake is not only the drinking water source of Shanghai people, but also plays the role in transportation, agricultural farm irrigation, aquiculture, impounding control and flood discharge and so on.

1.2 Natural Conditions

The water source area of Dalian Lake includes the fresh water lake, marsh, river network, shallow pond, fishpond and waterborne forest. The water source area has the dense water networks and its water area is 55.7 % of the total planned area. In this zone there is the largest *taxodium ascendens* forest; it is the largest one existing in

Fig. 1.2 Tai Lake water system

Shanghai area currently; the *taxodium ascendens* forest covers a land of 83 mu and has almost 8,000 trees; and each of them grows tall and well. There is relatively small population in water source area of Dalian Lake, where the population per square kilometer is 520, which is one sixth of the average population of Shanghai City, and the per capita arable land is 1.21 mu, which is four times of that of farmers of Shanghai City. Dianshan Lake area in which the water source area of Dalian Lake is located has about 80 % of the wetland biological species of the fresh lake of Shanghai City. Due to the local biological resources are quite rich and the biological diversity is at a high level, the Dianshan lake area is always regarded as one of few rare wildlife habitats in Shanghai Area. The water source area of Dalian Lake is the land of Shanghai City for verdurization and is reserved the basic nature of the farm land.

Dalian Lake belongs to the Dianshan Lake water system; its natural conditions are basically consistent with that of Dianshan Lake. Dalian Lake looks like a calabash and covers a land of 4.6 km^2, where the perennial average water depth is about 2 m. Dalian Lake connects with Lanlu Port; in ancient times it was a land.

Dalian Lake area belongs to the North Asian tropical monsoons climate; where it is mild, moist and there are four distinct seasons, sufficient sunlight, rich rainfall and long frost-free period. The yearly average sunshine duration of the lake zone is 1,930 h,

Fig. 1.3 Shanghai water source protection zone

and the yearly sunshine percentage is 44 %. The time to have the maximum sunshine duration is August, where it is 237 h averagely and its sunshine percentage is 58 %; among the whole winter, the time to have the minimal sunshine duration is February, where it is 112 h averagely and its sunshine percentage is 36 %. According to the record, the sunshine duration in the year of 1967 was 2,277 h, and it was the year having the maximum sunshine duration in past 50 years; the sunshine duration in the year of 1948 was 1,459 h, and it was the year having the minimal sunshine duration in past 50 years; the difference between both of them is 818 h.

The yearly average temperature of Dalian Lake is 15.8 °C; January is the coldest time and the monthly average temperature is 3.6 °C; the extreme minimal temperature (On January 19, 1893) was −12.1 °C; the hottest month is July and the monthly average temperature is 27.8 °C; the extreme hottest temperature was 40.2 °C (on July 12, 1934).

The annual mean rainfall in Dalian Lake area is 1,149 mm; the rainfall in the period from May to September is generally over 100 mm/month; the time to occur the maximum rainfall is June and it is 174 mm/month averagely; the time to occur the minimal rainfall is December and it is 37.5 mm/month averagely. In 1985, a year in which the maximum rainfall occurred, the rainfall was 1,673 mm; in 1982, a year in which the minimal rainfall occurred, the rainfall was 709.2 mm. There are 132 yearly mean rainfall days locally.

For Dalian Lake area, the S-oriented wind prevails in summer; in the period from June to August, the wind direction frequency is 53 %; the N-oriented wind prevails in winter; in the period from December to February of next year, the wind direction frequency is 54 %; the spring and the autumn are the two seasons in which the south and the north wind alternates; in spring the E-SE wind prevails; but in autumn the E-NE wind prevails; the average wind speed is 3.1 m/s.

In Dalian Lake area, the yearly mean frosty days are 37.8 days; where the time to occur the frost is in the middle 10 days of November averagely; the time on which the frost ends is in the last 10 days of March of next year averagely; the time to occur the frost at the earliest was October 22, 1979; the time on which the latest frost occurred was April 23, 1959; the mean period between the earliest frost day and the latest frost day is 126 days. In a whole year, the month in which the frost day occurs maximally is January and it is 11 days averagely; following it is December and it is 10.3 days averagely; in February and March, the average frosty days are 4–8 days.

In the period from 1875 to 1990, there were totally 300 typhoons happened in Dalian Lake area; its yearly mean typhoon frequency was 2.6; the maximum typhoon frequency was 7 and the minimal typhoon frequency was 0; among which the typhoon being accompanied with the high wind over 10 scale and the typhoon being accompanied with the rainstorm was respectively 21 and 24 % of the total frequencies. Generally the typhoon occurs in the period from May to November; among which it happens maximally in July, August and September, which is 83–89 % of a whole year. In August the typhoon happens maximally, which accounts for 36–39 % of a whole year. Averagely the influence caused by a typhoon lasts for 2.6 days, 8 days maximally and 1 day minimally; over half typhoons last for 1–2 days and the typhoon lasting for 5 days or more accounts for 11 %.

The cold wave occurs 3.5 frequencies averagely in a year in Dalian Lake water area; it occurs frequency up to eight times, at least for 0. The maximum cold wave occurred in 1960s, where it occurred frequency is 4.9 times per year on average, in 1970s, it occurred frequency is 3.1 times per year on average; in 1980s, it occurred minimally and the average frequency was 2.4. The earliest cold wave occurs in the third 10 days of October and the latest one occurs in the middle 10 days of April; the cold wave occurs maximally in March, following it is December.

Influence of Dalian Lake water area cyclone occurrence frequency is 33.1 times per year on average, up to 44, at least for 21 times. Normally the cyclone occurs in May and June maximally; in summer it reduces gradually and becomes minimal in August; among which the cyclone that may develop is 34 %. In a year the cyclone rainstorm occurs frequency is 2.5 times, the large rainstorm with the rainfall over 100 mm is 12 %; most of the cyclone rainstorm are the partial rainstorms, and the cyclone rainstorm occurs maximally in June; and the large rainstorm occurs in June mainly too; averagely there is a typhoon rainstorm in a year; in the maximum year, the typhoon rainstorm occurs frequency four times; but in the minimal year, it occurs never.

The yearly mean high water level of Dalian Lake is 2.68 m, the yearly mean low water level is 2.03 m; the historical highest water level is 4.04 m, which occurred on July 2, 1999; the historical minimal water level is 1.04 m, which occurred on January 9, 1956 (Data source: Maodian Station).

In accordance with the data of Qingpu Sub-centre (Qingpu District Hydrology Survey Team) of Shanghai Water Environmental Monitoring Center in the year of 2005, the ammonia nitrogen, total phosphor, chemical oxygen demand and other indexes of Dalian Lake water area, in view of the evaluation based on *Environmental Quality Standard for Surface Water* (GB3838-2002), are mainly centralized in Category V and inferior Category V; where the integrated evaluation of water quality is Category III–V. Among which the water quality of Taipu River and Lanlu Port belongs to Category III basically; the stage performance is Category II; the index of total phosphor and total nitrogen of Dianshan Lake zone belongs to interior Category V basically. The water quality in the west region of Qingpu is better than that of the water in its middle region and eastern region; and the water quality in the non-flood season is roughly better than the one in the flood season.

1.3 Social Economic Conditions

Dalian Lake water source area involves nine natural villages of two towns (namely Jinze Town and Zhujiajiao Town); there are 7,504 populations (2,650 households) in the zone currently, and the local planning area is 14.6 km^2.

From Fig. 1.4 and Table 1.1, it can be seen there are large village density and population density in Dalian Lake water source area; where the population density is 514 people/km^2. Despite of the water factors, for instance, lakes, ponds, rivers etc., the density of population on land is 1,160 people/km^2; where the per capita agricultural acreage is 0.88 mu approximately.

In view of the analysis described by Table 1.2, it can be seen the economic development level of Dalian Lake water source area is rather low comparing with the other regions of Shanghai City, and the percentage of the local three industries is inharmonious; where its regional development should focus on "the one to improve the percentage of the tertiary industry approximately, limit the secondary industry and transform the primary industry". The ecological restoration project of Dalian Lake water source area should provide sound opportunity for the local area for adjusting its industrial structure.

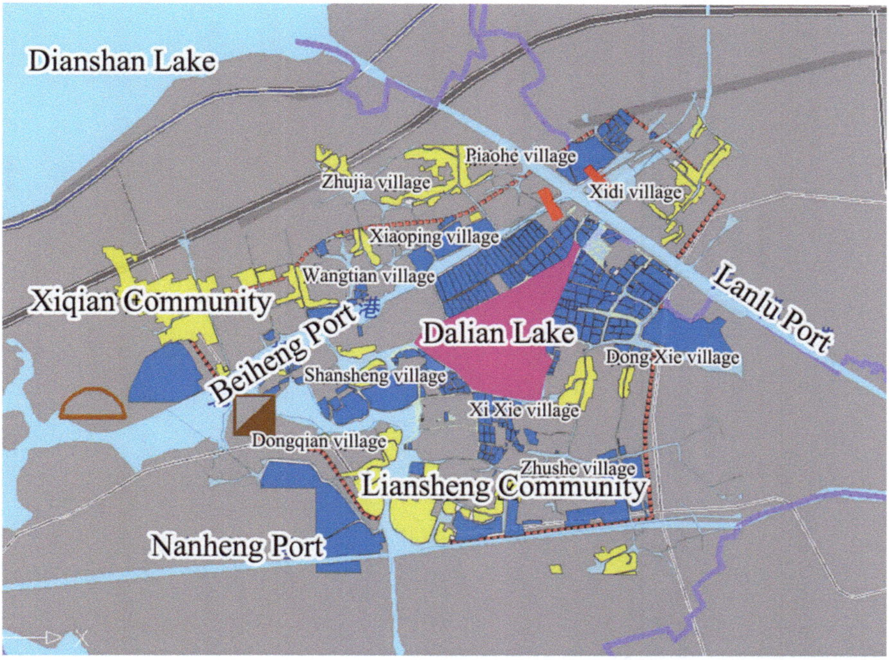

Fig. 1.4 Villages in Dalian Lake water source area

Table 1.1 Land type of Dalian Lake area

Land type	Area (ha)	Percentage (%)
Water system	471.58	32.3
Artificial wetland	341.64	23.4
Arable land	442.38	30.3
Non-farm land	204.4	14.0
Scope of phase II	1,460.0	100.0

Table 1.2 Economic, social and environmental development index of Dalian Lake water source area

Index		Unit	Status quo	Recent objective	Forward objective	Level of moderately developed country	Basic modern index system of Chinese Academy of Sciences	National standard for ecological city construction
Economic development index	Per capita GDP	US dollars	1,070	4,000	20,000	10,000	>5,000	≥4,000
	Percentage of the tertiary industry to GDP	%	5.7	45	65	65	60	≥45
	Percentage of R&D input to GDP	%	0.5	2	3	2.2–3	>3	
	Energy consumption per GDP	Tonnage standard coal/RMB10,000	1.0	0.85	0.5			≤1.4
Social development index	Average life expectancy	Year old	76.3	77	80	77.5		
	Percentage of active junior college students	%	1.5	10	50			≥30
	Percentage of education expenditure to GDP	%	—	3.5	5	5.4	>5	
	Per capita living space	m^2	—	22	35	35	>30	
	Doctor number/1,000 persons	Person	2.0	>5	6.5	5	>5	
	Social security coverage ratio	%	—	90	100	90–100		
	Registered unemployment rate of residents	%	—	3	3	2–6	4	
	Engel coefficient	%	48.1	32	30	30		<40

Environmental development index	Forest coverage	%	10	26	55	30		≥15
	Environmental quality of surface water	Category	III–IV up to standard roughly	II–III	II–III up to standard fully			Up to standard and without exceeding Category IV
	Atmospheric environmental quality	Grade	1–2	1–2	1	1	≥2	≥Class 2, 330 days/year
	Centralized processing ratio of domestic sewage from villages and towns	%	—	30	100		>60	≥70
	Hazard free treatment ratio of house refuses from villages and towns	%	—	60	100		>80	100
	Cycling utilization rate of industrial water	%	8.33	25	60			≥50
	Per capita area for public green space	m^2	6.3	15	25		>10	≥11

Bibliography

Chen, D. C. (2001). Eco-economics system and protecting environment form pollution in Qingpu District [J]. *Ecological Economy, 12*, 74–75 (in Chinese).
Ruan, R. L., Tu, H. M., & Wang, Y. (1997). Research of water quality monitoring and eutrophication strategies of Dianshan Lake [J]. *Water Conservancy of Shanghai, 48*(3), 35–38 (in Chinese).
Wang, S. M., & Dou, H. S. (1998). *The lakes of China* [M] (p. 55). Beijing: Science Press (in Chinese).

Chapter 2
Regional Evolution of Dalian Lake

2.1 Regional Wetland Type

As same as the forest, grassland, desert and ocean, the wetland is an important integral part of global ecological environment and an ecological system with particular hydrologic, soil and biological features. Generally the wetland refers to the natural transitional zone from water to land. The word of "wetland" was firstly mentioned by U.S. Fish and Wildlife Service in the publications titled by *No.* 39 *Notice* in 1956. In which the wetland was defined as the lowland covered by the intermittent or the permanent shallow water layer. Such definition satisfies with the needs of wetland administrators and wetland scientists finitely, and thence it is still adopted and used frequently by wetland scientists and wetland administrators nowadays. In 1979, Zoltai, in a discussion meeting of Canadian National Wetland Working Party, defined the wetland into the one as follows, *i.e.*: "the land, where the wet soil dominates and the aquatic plants grow, but the water level approaches to or exceeds mineral soil in the majority days in the unfrozen seasons". The definition made by the Canadian for the word of "wetland" limits the hydrologic conditions and wet soil conditions more concretely. Upon years' investigation, the wetland scientists of U.S. Fish and Wildlife Service defined the wetland as the one as follows, *i.e.*: "the wetland is a transitional zone of the terrestrial system and the aquatic system; on these lands, the water level is normally on or approaches to the earth's surface, or is covered by the shallow water".

In view of the connotation of the wetland, it comprises three features as follows, *i.e.*:

1. The dominant aquatic plants grow on the land periodically at least;
2. The watertight aqueous soil in the matrix dominates;
3. The non-soil matrix is sometimes saturated by water or covered by shallow water in growth season.

S. An and L. Wang, *Wetland Restoration: Shanghai Dalian Lake Project*,
Springer Environmental Science and Engineering, DOI 10.1007/978-3-642-54230-5_2,

For the definition of the wetland, it is usually adopted the definition in Ramsar Wetland Convention currently, *i.e.*: "the wetland refers to the natural, artificial, permanent or temporary marsh, peat land or water area, whichever it is static, flows, fresh water, brackish water, or salt water, including the water area with the water depth not more than 6 m in the low tide". In this book the lake wetland definition in the Ramsar Wetland Convention is adopted.

The main factors affecting the wetland functions include the hydrologic factor (rainfall inflow process, climate before and after rainfall) and the hydrodynamic force factor (movement status and dwelling time of water flow in wetland), and the other hydrologic factors, for instance, non-point source pollutant, retention rate of runoff and runoff process, dwelling time, water depth, etc. have close relation; however, the dwelling time of the runoff in the wetland system is controlled by the runoff inflow process. The main mechanism of the wetland system to control the non-point source pollution is to increase the runoff infiltration, delay runoff flow, increase dwelling time, etc., and then make pollutants detain in the system to decompose and transform, and finally reduce the pollutant loads from going into the downstream water body; Meanwhile, the rainfall runoff intercepted shall be recycled, which can improve the utilization ratio of water sources and relieve the scarce water source.

When designing the wetland system for controlling the non-point source pollution, it is needed to consider the characteristics of rainfall runoff (rainfall strength, rainfall duration, rainfall time interval, etc.), wetland dewatering time, drought tolerance of plants, transpiration and other factors so as to make the structure and functions of wetland fit to the randomness and rapid change of rainfall. The load of unit area affects the degradation efficiency of the wetland on pollutants; the improper design can make wetland system unworkable; in other words, the primary issue of the design work is to determine the pollution load of wetland area (currently it is needed to determine the pollutant reduction objective of wetland in accordance with the pollution source investigation result, and then obtain the pollution load of wetland area). In the wetland, the degradation extent of pollutant is related to its dwelling time; whereas the dwelling time is inversely proportional to the unit area's load of wetland.

Dianshan Lake evolves from the ancient lagoon in the Holocene period; the whole lake area is the plain water network zone formed under the fluvial outwash of the Yangtze River and the depositing effect of the ancient lagoon in the Quaternary period. According to the record of "Qingpu county annals", the shore of Dianshan Lake in the Song Dynasty was about 100 km long; but now the whole lake shore is about 62.3 km long. In Dalian Lake Demonstration Zone, there are mainly the aquatic forest, farming point, watercourse and *Phragmites communis Trin* marsh etc.; where Douji Port and Dalian River pass through the demonstration zone and connect with Lanlu Port, which is about 40 ha. Dalian Lake is an integral part of Dianshan Lake water area; its formation reason is rightly as same as that of Dianshan Lake. According to the wetland category, Dalian Lake wetland belongs to the mixed wetland of the natural lake wetland and the artificial wetland caused by fishponds.

Meanwhile, Due to the river courses in Dalian Lake wetland are open to the outer area, the biological species includes not only the conventional sedentary species, but also the migration-oriented species, as well as the external species; the hybrid species causes the local original species to change from time to time.

2.2 Regional Wetland Biology

2.2.1 Plankton

In the period from May to August 2008, Nanjing University, Nanjing Agricultural University and Shanghai Ocean University and other units established an investigation party to investigate the biological resources in Dalian Lake water area; the collection, stabilization and number of phytoplankton sample was done by referring to "*Lake Eutrophication Investigation Code*". Additionally 1 L water sample was taken, and the 25 mL methanal and 15 mL iodine solution were adopted to stabilize the water sample; afterwards the water sample was taken back to the laboratory and kept still for 48 h; following the clear solution was absorbed and the rest was concentrated to 50 mL; after shaking sufficiently, 0.1 mL sample was taken by the pipette to make the category appraisal and cell counting under the optical microscope; where the references for such category appraisal were the documents prepared by Han Maosen et al. (1980), every sample was counted two pieces repeatedly, and the sample with the tolerance exceeding 15 % was re-counted one piece additionally. For the investigation scope and sampling point, see Fig. 2.1 as follows.

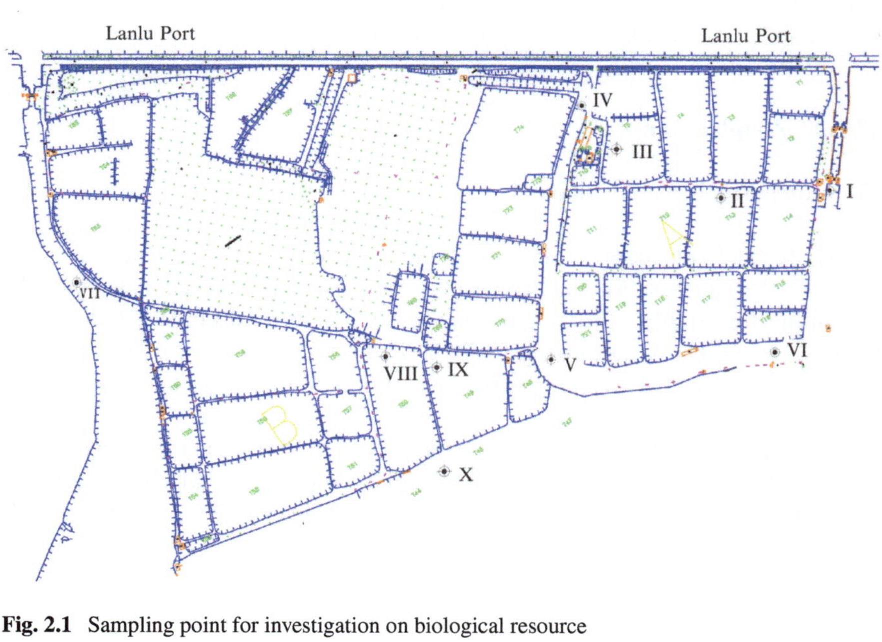

Fig. 2.1 Sampling point for investigation on biological resource

2.2.1.1 Phytoplankton

Kinds of Phytoplankton

Upon the investigation, totally 86 kinds of algae were found; all of them belong to 7 phyla and 48 genuses (see appendix Table A.1). Among which there were 11 genuses and 17 kinds of cyanophyta, accounting for 22.9 % of the total phyla and 19.8 % of the total kinds; 1 genus and 14 kinds of *Xanthophyta*, accounting for 2.2 % of the total phyla and 1.3 % of total kinds; 1 genus and 1 kind of *Pyrroptata*, accounting for 2.1 % of total genuses and 1.2 % of total kinds; 1 genus and 2 kinds of *Cryptophyta*, accounting for 2.1 % of the total genuses and 2.3 % of total kinds; 2 genuses and 6 kinds of *Euglenophyta*, accounting for 4.2 % of the total genuses and 7.0 % of total kinds; 7 genuses and 12 kinds of *Bacilariophyta*, accounting for 14.6 % of total genuses and 14.0 % of total kinds; 25 genuses and 47 kinds of *Chlorophyta*, accounting for 52.1 % of total genuses and 54.7 % of total kinds. Among which the kind of VII algae from the sampling point of the external river course was minimal, but the kind of II algae from the sampling point of pond was maximal.

Dominant Species of Phytoplankton

In view of the density and the distribution of phytoplankton, the dominant species of the phytoplankton in Dalian Lake water source area are: *Phormidium sp.* and *Merismopedia tenuissim of Cyanophyta, Cryptomonas erosa of Cryptophyta, Synedra ulna of Bacillariophyt*a, *Chlorella vulgaris*, *Tetraëdro trigonum*, *Chlamydomona simplex* and *Scenedesmus quadricauda of Chlorophyta.*

2.2.1.2 Zooplankton

Kind Composition and Distribution

Totally 24 kinds of zooplankton were found in Dalian Lake water source area, belong to 15 genuses. Among which the rotifer is of 6 genuses and 13 kinds, accounting for 40.0 % of total genuses and 54.2 % of total kinds; *Cladocerans* 4 genuses and 6 kinds, accounting for 26.7 % of total genuses and 25.0 % of total kinds; copepods 5 genuses and 5 kinds, accounting for 33.3 % of total genuses and 20.8 % of total kinds.

The kinds of zooplankton found in the samples from all river courses and marshes were found maximally, but it is rather less in the samples from all sampling points of ponds. No rotifer was found in the samples from the sampling points VI, VII, VIII and IX.

Dominant Species of Zooplankton

In view of the distribution and the density, the dominant species of the zooplankton in Dalian Lake water source area are *Brachionus calyciflorus*, *Diaphanosoma brachyurum* and *Cyclops vicinus*.

2.2.2 *Aquatic Vascular Bundle Plants*

There were 10 vascular bundle plants in Dalian Lake water source area, including 3 floating plants, 3 submerged plants, 3 emergent aquatic plants and 1 floating-leaved plant. The *Phragmites communis* and the *Zizania caduciflora* are full of the coastal areas of river courses, scattered with *Paspalum paspaloides*; the water surface of river courses is full of *Spirodela polyrhiza,* with additional piles of *Hydrocharis dubia* (Blume) Backer and *Alternanthera philoxeroides*; the submerged plant includes: *Cabomba caroliniana*, *Ceratophyllum demersum*, and *Potamogeton malaianus* Miq.; where the dominant plants in the coastal areas of ponds include the *Phragmites communis* and the *Paspalum paspaloides*, and the one on the water surface includes: *Spirodela polyrhiza, Hydrocharis dubia*, *Alternanthera philoxeroides* and *Trapa natans.*

2.2.2.1 Floating Plant

Spirodela polyrhiza, *Lemnaceae*, *Spirodela*; *Hydrocharis dubia*, Hydrocharitaceae, *Hydrocharis*; *Alternanthera philoxeroide*s, Amaranthaceae, *Alternanthera*.

2.2.2.2 Submerged Plant

Ceratophyllum demersum, Ceratophyllaceae, Ceratophyllum; Potamogeton malaianus, Potamogetonaceae, Potamogeton; Vallisneria natans, Hydrocharitaceae, Vallisneria.

2.2.2.3 Emergent Aquatic Plant

Phragmites communis, Gramineae, Phragmites; Zizania caduciflora, Gramineae, Zizania; Paspalum paspaloides , Gramineae, Paspalum.

2.2.2.4 Floating-Leaved Plant

Trapa natans, *Trapaceae*, *Trapa*.

Table 2.1 Kind composition and distribution of zoobenthos in every sampling point

Kind		Sampling point							
		I	IV	V	VI	VII	VIII	IX	X
Mollusca									
Bellamya	*B. aeruginose*		+			+			
	B. purificata	+		+	+		+		+
	B. purificata	+		+	+	+			+
Angulyagra	*A. polyzonata*	+			+				
Alocinma	*A. longicornis*	+	+	+	+	+	+		
Parafossarulus	*P. sinensis*	+				+			
Limnoperna	*L. lacutris*	+			+				
Unio	*U. douglasiae*	+				+			
Anodonat	*A. fluminea*					+			
Annelida									
Limnodrilus	*L. claparedeianus*		+		+			+	
	L. hoffmeisteri		+	+			+	+	
Branchiura	*B. sowerbyi*	+						+	
Tubifex	*T. tubifex*								
Arthropoda									
Chironomus	*C. flaviplumus*					+		+	+
Dicrotendipes	*D. pelochloris*							+	
Statistics		8	4	4	6	7	3	5	3

2.2.3 Zoobenthos

2.2.3.1 Kind Composition and Distribution of Zoobenthos

The substrate of the sampling point II and III in Dalian Lake water source area pond is the hard sediment, where no zoobenthos were found. As for the other sampling points, 15 zoobenthos were found upon the appraisal, belonging to 3 phyla and 12 genuses (Table 2.1); among which, the mollusca is of 7 phyla and 9 kinds; oligochaeta is of 3 phyla and 4 kinds; Chironomidae is of 2 phyla and 2 kinds, accounting for 60 %, 26.7 % and 13.3 % of the total kinds respectively.

2.2.3.2 Dominant Zoobenthos in Dalian Lake

The dominant zoobenthos in Dalian Lake water source area includes: *B. purificata*, *B. purificat* and *A. longicornis.*

2.2.4 *Fishes*

Dalian Lake has the characteristics of the fish compound ecosystem of Dianshan Lake water system, dominating by cyprinid fish; following it is the sedentary small freshwater fish, estuary-oriented rich saltwater fish and migratory fish; where the main fishes include: *Coilia nasus*, *C. nasus taihuensis*, *Neosalanx tangkehkeii taihuensis Chen*, common carp, crucian carp, gurnard, *Erythroculter mongolicus*, *Topmouth culter*, *Erythroculter illshaeformis*, *Erythroculter dabryi*, *Paracanthobrama guichenoti*, *Hemibarbus maculatus*, *Elopichthys bambusa*, *Saurogobio dabryi*, *Hemicculter Leuciclus*, various *Rhodeus sinensis Gunther*, catfish, eel, mullet, *Siniperca chuatsi*, *Cymoglossus robustus*, *Channa argus*, *Takifugu obscurus*, and other breeding fishes, such as grass crap, black carp, silver carp, bighead carp, Bluntnose black bream, carp and so on. Upon the investigation on Dalian Lake water source area, the fishes under 5 orders, 9 families and 24 species were found.

2.2.5 *Birds*

There are flourishing vegetations, excellent habitat and special rich bird resources in Dianshan Lake area. According to the investigation, the birds under 17 orders, 48 families and 183 species were found, including 1 bird species under the first-grade State protection (*Mergus squamatus*), 39 bird species under the second-grade State protection, 1 bird species under the Chinese endemic species (*Paradoxornis heudei*), 8 bird species under Chinese White Paper for Endangered Animals and 33 bird species under *Convention on International Trade in Endangered Species of Wild Fauna and Flora* (CITES).

2.2.6 *Animals*

The animals recorded in Dianshan Lake area involve in 9 orders and 34 species, including 3 animals under the second-grade State protection (Common Otter, rasse, zibet), and 5 animals under CITES.

2.2.7 *Endangered Rare Species*

According to the years' investigation and the historical records, there are 40 mammalian species, 424 bird species (including the subspecies), 32 Reptiles species, 14 amphibians species and 250 fish species in Shanghai area. Among which there are 9 wildlife animal species under the first-grade state protection, for instance, hooded crane, Ciconia boyciana, Chinese sturgeon, etc.; 62 wildlife animal species under the second-grade State protection, for instance, black-faced spoonbill, grey crane, *Cygnus columbianus*, *Platalea leucorodia*, tiger frog, sea turtle, etc.; 56 species in the appendix to *Convention on International Trade in Endangered Species of Wild Fauna and Flora* (CITES), 185 birds species under *Agreement of the People's Republic of China and Japanese on the Protection of Migratory Birds and their habitats*, 86 bird species under *Agreement of the People's Republic of China and Australia on the Protection of Migratory Birds and their habitats*, and 46 wildlife animal species under the key protection of Shanghai City, for instance, leopard cat, *rrinaceus earopaeus*, hog-nosed badger, magpie, *zaocys dhumnade*, etc.

2.2.8 *Invasive Species*

The invasive species in Dalian Lake water source area includes *Cabomba carolimana, Eichhirnia crasslpes* and *Ampullaria gigas* Spix, etc.

2.3 Regional Wetland Evolution

Dianshan Lake water area evolves from the ancient lagoon in the Holocene period; the whole lake area is the plain water network zone formed under the fluvial outwash of the Yangtze River and the depositing effect of the ancient lagoon in the Quaternary period. According to the record of *Qingpu county annals*, the bank of Dianshan Lake in the Song Dynasty was about 100 km long; but now the whole lake bank is about 62.3 km long. Dalian Lake is an integral part of Dianshan Lake water area; its formation reason is rightly as same as that of Dianshan Lake. According to the wetland category, Dalian Lake wetland belongs to the mixed wetland of the natural lake wetland and the artificial wetland caused by fishponds.

Due to the river courses in Dalian Lake wetland are open to the outer area, the biological species includes not only the conventional sedentary species, but also the migration-oriented species, as well as the external species; the hybrid species causes the local original species to lose and change from time to time.

Since the beginning of the twentieth century, as the local population increases hugely and the frequent actions to reclaim farmland from lakes, the lake area of Dalian Lake has reduced sharply. The water area of Dalian Lake in 1980s was 1.474 km^2, but now it is only 0.9 km^2. Additionally, due to river courses are silted seriously and the lake is netted to breed fishes, which causes poor flow and high density of suspended matters to Dalian Lake water area, and makes the water of whole lake area eutrophicated; which affects the native species for survival and the migrant fishes for activity seriously. Currently plenty of important species in Dalian Lake area have already extinguished, and the Gorgon fruit, Brasenia schreberi and myxiocyprinus asiaticus are at the risk of disappearing too.

2.3.1 *Kinds of Phytoplankton and Evolution of Dominant Species*

In recent 50 years, the blue and green algae have been growing rapidly in Dalian Lake water source area. According to the data, the phytoplankton quantity in Dianshan Lake in 1959 was not beyond 2,000/L, but in 1991, it was 9×10^5/L; however in 2000, it exceeded 10×10^7/L; in the year of 2008, the phytoplankton quantity in Dianshan Lake was 2.04×10^7/L; but a downward trend in the percentage of the algae fitting for aquatic economic (Table 2.2).

2.3.2 *Kinds and Quantity Evolution of Zooplankton*

The zooplankton is the first secondary productivity of lake and the important nature bait for fishes. The kinds and changes of zooplankton reveal the influence of lake eutrophication to lake ecology to some extent (Table 2.3).

In view of the historical investigation results, it can be seen the zooplankton in Dianshan Lake water area is mainly dominated by protozoa, rotifera, cladocera animals and copepoda animals under the crustacean category. In view of the biological number, the rotifera and the protozoa are the main zooplankton. The number of the rotifera is an important indicator species for the eutrophication extent of water. In view of Table 2.4 as follows, it can be seen the number of the rotifera in 2000 increased highly comparing with the one in 1982; which shows the water eutrophication intensifies to some extent.

2.3.3 *Type, Quantity and Distribution Evolution of Aquatic Vascular Bundle Plants*

The aquatic vascular bundle plants are the higher aquatic plants with the vascular bundle structure. According to the ecological features, the aquatic vascular bundle plants can be divided into four categories, namely emergent aquatic plant, floating-leaved plant, floating plant and submerged plant.

In accordance with the two investigations made by East China Normal University respectively in the period of 1983–1985 and 1987–1988, the aquatic vascular bundle plants under 23 families and 26 species were found in Dianshan Lake water area. The two investigation results had no big difference, but no *Nymphoides peltatum* (Gmel.) O. Kuntze, *Pseudobulbus Cremastrae Seu PLeiones, Nymphaea alba, Spirodela polyrr-hiza Schleid* were found by the investigation made in 1987–1988, however the *Najas marina, Najas minor-Caulina fragilis, Potamogeton pectinatus* and *Alternanthera philoxeroides* were additionally found in the investigation sample. The emergent aquatic plants in Dianshan Lake water area was absolutely

Table 2.2 Evolution of phytoplankton in Dianshan Lake water area

Year	Species/quantity								
	Cyanophyta	*Chlorophyta*	*Bacillariophyta*	*Cryptophyta*	*Pyrrophyta*	*Euglenophyta*	*Xanthophyta*	*Chrysophyta*	*Euglenophyta*
1982	13	28	26		3			1	3
1985–1986	6	14	7	2		2		2	
1987–1988	11	37	15	3	2	3	3	6	
2000	12	25	12	1	3	2	2	1	
2008	11	25	7	1	1	2	1		

Table 2.3 Kinds of zooplankton and evolution of their dominant species in Dianshan Lake water area

Investigation time (year)	Category	*Protozoa*	*Rotifera*	*Cladocera*	*Copepoda*
1982	Species type	28 genera 28 species	27 genera 51 species	14 genera 18 species	9 genera 11 species
	Dominant species	*Difflugia, Strombidium, Tintinnopsis*	*Diaphanosoma spp.*	*Limnoithona sinensis, Sinocalanus*	
1987–1988	Species type	20 genera 23 species	12 genera 24 species	9 genera 21 species	10 genera 12 species
	Dominant species	*Trinema lineare*	*Brachionus calyciflorus Pallas, Keratella quadrata, Polyarthra trigla*	*Diaphanosoma brachyurum, Bosminopsis deitersi*	*Limnoithona sinensis, Mesocyclops leuckarti, Schmackeria forbesi, Sinocalanus dorrii*
2000	Dominant species	*Epistylis lacustris*	*Brachionus plicatilis, Brachionus diversicornis, Asplachna priodonta Gosse, Trichocerca sp., Keratella*	*Diaphanosoma spp., Daphnia pulex, Moina, Bosmina, Chydorus sphaericus*	*Calanus, Cyclops, Canthocamptu spp.*
2008	Dominant species		*Brachionus calyciflorus Pallas*	Diaphanosoma brachyurum	Cyclops vicinus

Table 2.4 Comparison on zooplankton biomass in Dianshan Lake water area

	Biomass/(piece/L)			
Year	*Protozoa*	*Rotifera*	*Cladocera*	*Copepoda*
1982	754–1,320	81–400	24–56.1	18.5–40.56
1987–1988	1,400	783	60	140
2000	400	1,160	109	117
2008		710	370	500

dominated by the *Phragmites communis*; which was scattered in the south of Dianshan Lake and the lakeside area in Shangta Town; the floating-leaved plant in Dianshan Lake water area was the *Trapa*; the floating plants were mostly distributed in the broad water area, mainly the *Alternanthera philoxeroides, duckweed, Azoila imbricate, Hydrocharis dubia, Spirodela polyrrhiza, Eichhornia crassipes, Nymphaea alba* etc.; the submerged plants included *Vallisneria natans, Hydrilla verticillata, Ceratophyllum demersum, Myriophyllum spicatum, Najas marina* and *Najas minor. Caulina fragilis*, etc. Upon the two investigations above, It was found the kinds of the submerged plant different hugely each other and the submerged plant community changed hugely in the investigation in 1988, where a large area of *Myriophyllum verticillatum, Hydrilla verticillata, Najas marina* and other algae with less nutrient substances and hard to collect and utilize occurred; such algae have strong adaptability; if they grow into the main community of the submerged plants, it shall hugely change the ecological environment of the local water area.

In the year of 2000, Shanghai Ocean University Fishery College made an investigation in Dianshan Lake water area, where the aquatic plants under 2 families and 20 species were totally collected; among the collected aquatic plants, the *Vallisneria natans, Hydrilla verticillata, Ceratophyllum demersum* were the main species; the aquatic plants in the coastwise area were mainly the *Phragmites communis* and the *Zizania caduciflora;* the floating aquatic plants in water were mainly the *Eichhornia crassipes, Commom Duckweed, Salvinia natans*, etc.; the submerged plants were mainly the *Vallisneria natans*.

2.3.4 Type, Quantity Distribution and Evolution of Bentonic Organism

The bentonic organism is a group of organism living at the bottom of water. In August 1983, the Biology Department of East China Normal University made a qualitative and quantitative research on samples collected from 39 points (frequencies), where 21 mollusks, 8 crustaceans, 2 aquatic insects and 1 polychaetes animal and 1 oligochaeta animal were found. In the biocenosis the mollusk dominated, which accounted for 93.23 %; following it was the crustacean (3.69 %) and aquatic insect (0.9 %).

Again in the period from 1987 to 1988, East China Normal University, upon its another investigation, found the bentonic organism under 3 families and 6 classes, all of which belonged to 16 families 35 species; comparing with the investigation result in 1983, its community structure did not change hugely; where the mollusk dominated absolutely (90.9 %); however, the quantity of the Chironomid larvae and the oligochaeta organisms, which was stable originally, shown the increasing tendency; but the large mollusk animals, for instance, the number of Trigonioides, reduced and shown a tendency to become smaller; the number of the arthropods shown its reduction tendency too, even the Chinese soft-shell turtle was rare to see.

In 2000, ten zoobenthos were found in the samples collected by Shanghai Ocean University from the given collection points, which were mussel, corbicula fluminea, oncomelania, Bellamya, viviparid, Schistodesmus lampreyanus, Semisulcospira libertine, Chironomid larvae, clam worm, Bithynia tentaculata. Comparing with the investigation made by East China Normal University in 1983, the bentonic organism in Dianshan Lake water area, whichever it is the quantity or the species, shown its reduction tendency; however, the quantity percentage of the Chironomid larvae and Tubificidae approached to 50 %; which shown the pollution in Dianshan Lake water area became serious day after day.

2.3.5 Composition, Distribution Characteristics and Aquatic Product Types of Fishes in Dianshan Lake

The Dianshan Lake water area is an integral part of the Tai Lake water system, having the characteristics of the fish compound ecosystem of Tai Lake water system; where the cyprinid fish dominates; following it is the sedentary small freshwater fish, estuary-oriented rich saltwater fish and migratory fish. The main fish species in Dianshan Lake water area includes: *Coilia nasus*, *C. nasus taihuensis*, *Neosalanx tangkahkeii taihuensis Chen*, common carp, crucian carp, gurnard, *Erythroculter mongolicus*, *Topmouth culter*, *Erythroculter dabryi*, *Paracanthobrama guichenoti*, *Hemibarbus maculatus*, *Elopichthys bambusa*, *Saurogobio dabryi*, *Hemicculter Leuciclus*, various *Rhodeus sinensis Gunthers*, catfishs, eels, mullets, *Siniperca chuatsi*, *Cymoglossus robustus*, *Channa argus*, *Takifugu obscurus*, and other breeding fishes, such as grass crap, black carp, silver carp, bighead carp, Bluntnose black bream and mirror carp and so on. Upon the multiple investigations made by Shanghai Ocean University respectively in 1959, 1974, 1981–1982, 1982–1985 and 1987–1988, the fishes under 60 families 75 species, 47 families 61 species, 42 families 62 species, 44 families 55 species and 34 families 45 species were found accordingly. Upon the multiple investigations on the composition of fish fauna, the wildlife cyprinid fish were dominated in 1950s, but the breeding fish species were very rare; however, there were a quantity of estuary-oriented fishes and the migratory fishes, and some large yellowcheck carps; in 1970s, the migratory fishes and the estuary-oriented fishes reduced in Dianshan Lake water area, but the breeding fish species were increased rapidly. In view of the yield data obtained from the fishery service, the breeding fish

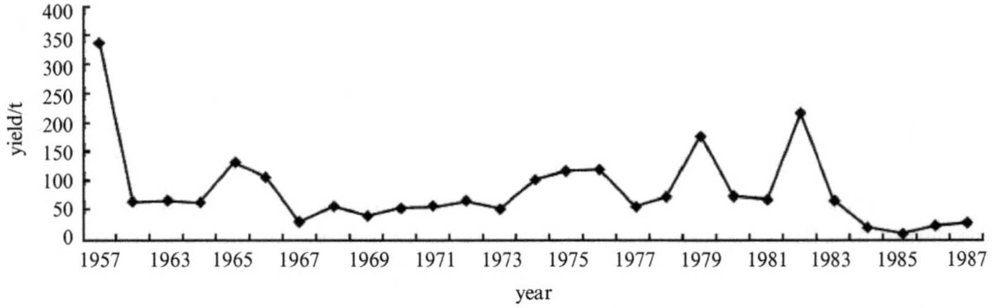

Fig. 2.2 Yearly crab yield in Dianshan Lake water area

species in Dianshan Lake water area increased to 50 % in 1980s from 5 % in 1950s and 30 % in 1970s.

In view of the multiple investigation results, the fish fauna in Dianshan Lake water area changed hugely in the past 30 years, and 30 fish species disappeared. The investigation result 2008 shown the fish species available in Dianshan Lake water area currently was only 40 % of the one in 1959, where 14 fish species were under the cyprinoid; most of fish species reduced were the migratory fishes and the semi-migratory fishes; the direct reason to cause such reduction is the dam structure constructed by human being, which possibly blocked the migration channels of fishes; the other possible reason is the unreasonable fishing operations. Figure 2.2 shows the Chinese Wollhandkrabbe yield in Dianshan Lake water area in the past 30 years (Data source: Qingpu District Aquatic Products Department).

Bibliography

An, S. Q. (2003). *Wetland ecological engineering: Optimization mode of wetland resource protection and utilization* [M]. Beijing: Chemical Industry Press (in Chinese).

Han, M. S. (1980). *Freshwater plankton atlas* [M]. Auburn: Agricultural Press (in Chinese).

Jin, X. C., & Tu, Q. Y. (1990). *Lake eutrophication investigation code* [M] (pp. 10–15). Beijing: China Environmental Science Press (in Chinese).

Yu, X. D., Luo, T. H., Wu, Y. M., & Zhou, H. Z. (2005). A large-scale pattern in species diversity of amphibians in the Yangtze River basin [J]. *Zoological Research, 26*(6), 565–579 (in Chinese).

Yue, F., Luo, Z. K., Wu, D., Pei, E. L., & Wang, T. H. (2010). Species composition and biodiversity of fish community in Dalian Lake, Shanghai [J]. *Zoological Research, 31*(6), 657–662 (in Chinese).

Zhang, D. G., & Yang, Z. F. (2006). Ecological and environmental succession strategies for the Dianshan Lake [J]. *Reservoir Fisheries, 26*(1), 61–63 (in Chinese).

Zhao, P., Yuan, X., Tang, S. X., & Wang, T. H. (2003). Species and habitat preference of water birds at the eastern end of Chongming Island (Shanghai) in winter [J]. *Zoological Research, 24*(5), 387–391 (in Chinese).

Zhou, F. X, & Chen, J. H. (2005). *Freshwater microorganism atlas* [M]. Beijing: Chemical Industry Press (in Chinese).

Chapter 3
Characteristics of Hydrologic Environment

3.1 Hydrologic Characteristics

For the Dalian Lake water system, where the Douji Port and the Dalian Port are connected with the Lanlu Port (Fig. 3.1), and the floodgate in the west side is the Lianhu floodgate, and the floodgate in the east is the Douji floodgate. The Lanlu Port in the north is the main connecting water course of Dianshan Lake and Huangpu River; the water flows from the west to the east and flows into the Huangpu River.

The flood period of Shanghai is the period from June to September. Generally the Douji Floodgate is closed in July and August; the time to close the floodgate is not beyond half a month generally. The water level in the river course of the Douji Floodgate is 2.84 m (Marked height of Wu Song); the water level in winter is 2.64 m. When the water level of the internal river course reached to 3.29 m, the water pump at the floodgate shall start working to pump water till the water level recovers to or beneath the warning water level of 3.14 m.

For Dalian Lake, the widest section of the river courses to its Douji Port and Dalian Port is about 10 m, but the narrowest section is only 5 or 6 m; where allows the local small fishing boats to access. The floodgate in the west is the Lianhu Floodgate, the floodgate in the east is the Douji Floodgate. Such two floodgates are open usually for fishing boats to access, but only closed when flood arrives; in such case, the water pumps at the floodgate start working to divert the excessive water in fishing ponds and the internal rivers to Lanlu Port to reduce their water level. The Lanlu Port is a main connecting water course to Dianshan Lake and Huangpu River; where water flows from the west to the east and goes into Huangpu River. The river course to the Lanlu Port is about 30 m wide, where allows the large boats to pass.

Lanlu Port is in the north of Dalian Lake water source area. As a main river course in Qingpu district, Lanlu Port is situated in the southeast of Dianshan Lake and between the large and small Lianhu Lake. Lanlu Port originates from Guanwang Temple in the lakeside of Dianshan Lake, connects with Mao River in the south, and extends to Dongfanghong Bridge of Qingfeng Highroad; Lanlu Port is a main tidewater river port of Dianshan Lake, but also one of main shipping lanes to

S. An and L. Wang, *Wetland Restoration: Shanghai Dalian Lake Project*,
Springer Environmental Science and Engineering, DOI 10.1007/978-3-642-54230-5_3,

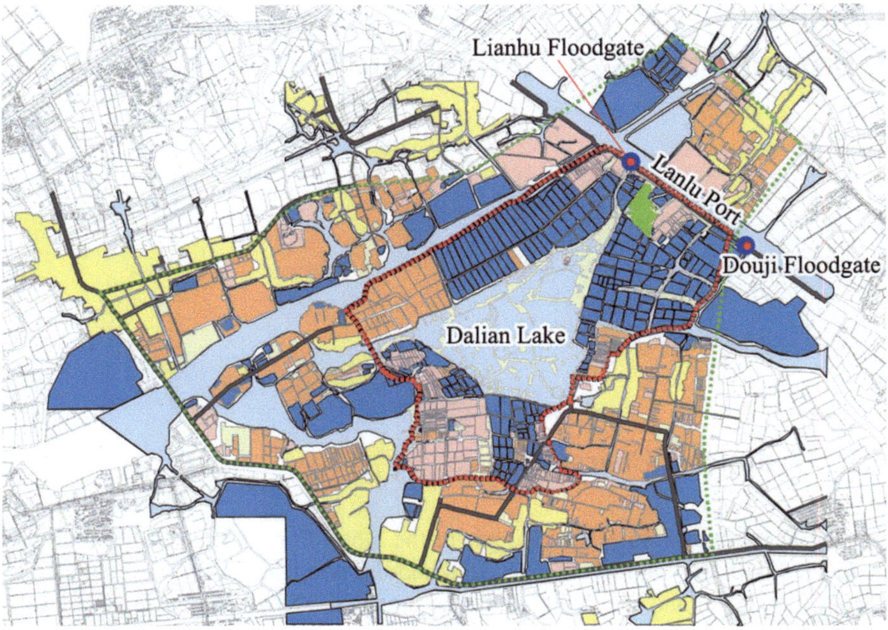

Fig. 3.1 Water system in the planned area of Dalian Lake

connect the Tai Lake watershed and Shanghai (Table 3.1). Lanlu Port is 8.7 km long, 30 m wide at the bottom, 3 m high at the bottom near to Dianfeng Peak, wide 80 m at the river surface and with the flow rate of 1 m/s. The river to Lanlu Port is wide and deep, where the water flows rapidly and the cross-section of river is 210 m^2. Lanlu Port is a Class V shipping lane and is one of main river reaches of Sushen Outer Port Line, allowing the ships of 200–300 t to pass and irrigate 20,000 m farmlands. There are three source flows in the upstream of Huangpu River, among which the W-N source flow is the main flow. Starting from Koudianfeng of Dianshan Lake, it is the Lanlu Port, with Mao River, Xie Pond, Sanjiaodu following; where Taipu River converges with Mao River; the W-N source flow carries the water coming from Tai Lake and Dianmao area of Jiangsu Province; after Taipu River is opened, it becomes the main water source of Huangpu River. From which it can be seen Lanlu Port is very important to the protection on the water quality of Huangpu River.

The average water depth of Dalian Lake in normal flow year is 2.0 m, the water surface area is 8.7 km^2 and the impoundage is 17,400,000 m^3; the average water depth in low flow year is 1.5 m, the water surface area is 6.5 km^2 (by 3/4), and the impoundage is 9,750,000 m^3; the evaporation and the water diversion are regarded as the water yield; the yearly mean precipitation in Dalian Lake area is 1,181 mm; the land evaporation is 764 mm, and the water surface evaporation is 842 mm.

Table 3.1 Data for tide level and precipitation of Lanlu Port in recent period (Maodian Monitoring Station)

Water level at 8:00 o'clock (m)	Maximal water level (m)	Time of tide	Precipitation (mm)	Date
2.83	2.99	19:05	—	2008-9-8
2.96	3.07	17:50	0.5	2008-9-7
3.07	3.10	19:05	14.0	2008-9-6
2.98	3.06	18:05	21.0	2008-9-6
2.88	3.01	5:20	—	2008-9-5
2.88	3.03	4:25	—	2008-9-4
2.85	3.04	3:50	—	2008-9-3
2.86	3.04	4:30	—	2008-9-2
2.81	3.05	4:35	1.0	2008-9-1
2.79	3.01	3:35	3.0	2008-8-31
2.75	2.98	3:40	1.0	2008-8-30
2.71	2.92	2:00	—	2008-8-29
2.90	3.10	3:40	6.5	2008-6-19
2.83	3.02	2:55	20.5	2008-6-18
2.77	3.00	2:40	—	2008-6-17
2.99	3.11	5:40	29.0	2008-6-22
2.90	3.05	5:25	—	2008-6-21
2.90	3.08	3:45	—	2008-6-20
Herein 3 days were the period in which the water level (tidal level) changed extremely in 2008				
3.29	3.37	5:45	32.5	2008-6-24
3.19	3.30	6:15	49.5	2008-6-23
2.60	2.79	23:40	—	2008-9-11

3.2 Physical and Chemical Characteristics of Water Environment

An overall investigation was made on the breeding ponds, river water and substrate of Dalian Lake water source area in 2008, where totally 14 water samples and 30 bottom sediment samples (including the blank) were made as per the grid layout sampling pattern (Fig. 3.2), all such samples covered ponds, rivers and aquatic forest and other zones in water source area.

For the water quality of Dalian Lake water source area, see Table 3.2 as follows. Rivers and aquatic forest were seriously polluted by nitrogen, where the total nitrogen went beyond 2.0 mg/L and thence the water quality belongs to the poor V grade (according to the water quality standard of surface water). Except some breeding ponds, most of them had the total nitrogen less than 1.5 mg/L. The ammonia nitrogen concentration is 0.43–0.79 mg/L; such concentration can not make breeding animals poisoned (1.0 mg/L), but it can affect their growth (<0.5 mg/L). Among the three kinds of nitrogen, the Nitrate Nitrogen is toxic minimally; when the Nitrate Nitrogen concentration in water body is over 1.0 mg/L, the penetration of aquatic

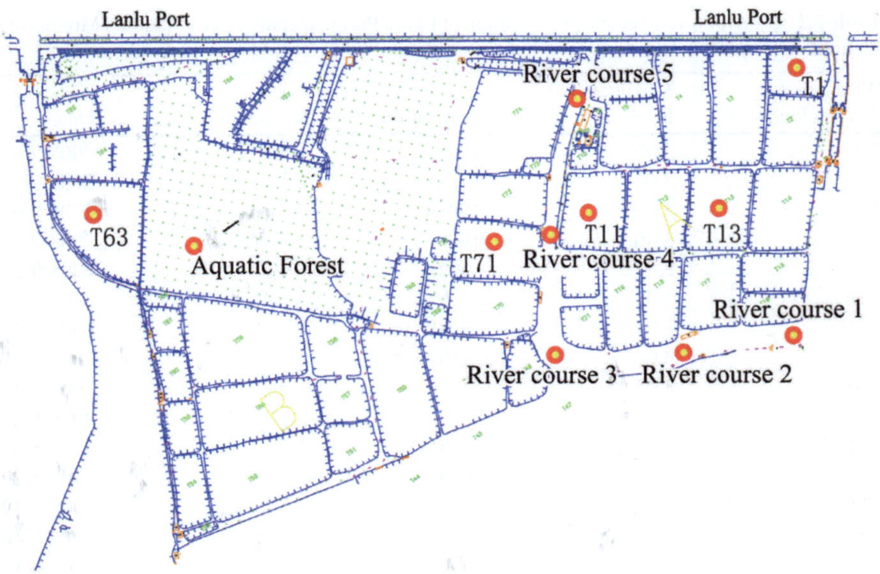

Fig. 3.2 Water quality and substrate quality sampling points in Dalian Lake water area

Table 3.2 Water quality of Dalian Lake water source area

Sampling point	Analysis item				
	Total nitrogen (mg/L)	Ammonia nitrogen (mg/L)	Nitrate nitrogen (mg/L)	Nitrite nitrogen (mg/L)	TSS (mg/L)
Aquatic forest	3.82	0.79	1.99	0.02	96
River course 3	2.06	0.65	2.17	0.04	135
River course 5	2.78	0.57	2.97	0.09	152
T1	2.57	0.43	2.2	0.01	182
T11	2.70	0.55	2.08	0.01	214
T13	2.15	0.52	1.52	0.01	154
T63	2.32	0.75	1.56	0.02	110

animals and the transport capacity of oxygen will be affected; when the Nitrite Nitrogen concentration is less than 0.1 mg/L, it shows the Nitrite Nitrogen concentration in water is normal. The total suspended solids in aquatic forest, river and breeding ponds are high, so any of them are not suitable for people or livestock for drink. Generally the total ammonium in the breeding ponds in summer and autumn is 0.5–4 mg/L, nitrite nitrogen 0.1–0.4 mg/L, nitrate nitrogen 0.1–2 mg/L; provided any of the limits above is exceeded, it shall mean the breeding water suffers with great pollution by organic pollutants. The water of the demonstration point of Dalian Lake is rather turbid, the aquatic forest and rivers are of eutrophication, and the

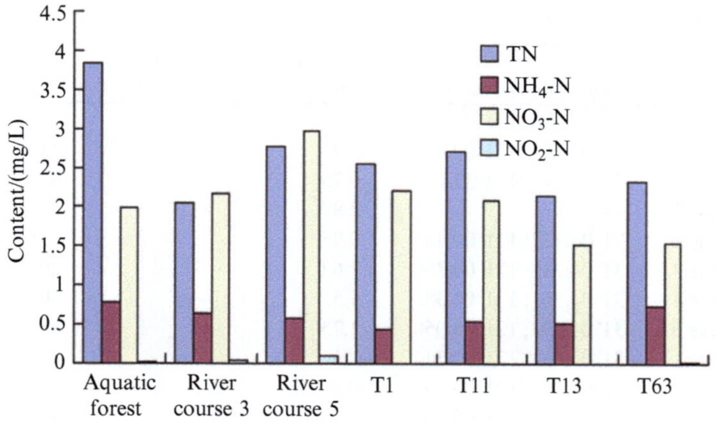

Fig. 3.3 Total nitrogen and three kinds of nitrogen in different water areas of Dalian Lake

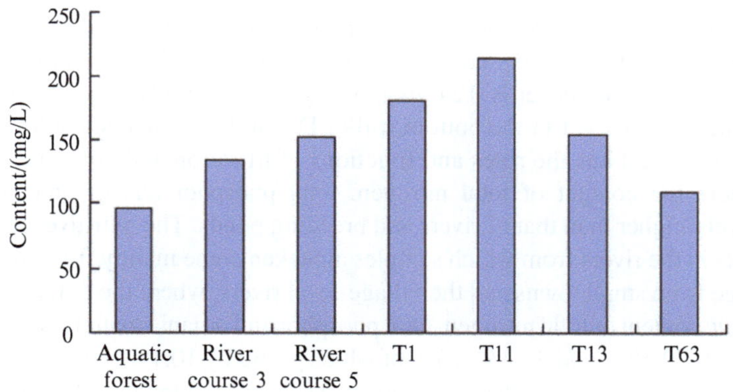

Fig. 3.4 Total suspended solids in different water areas of Dalian Lake

breeding ponds are under the risk of organic pollutants, so it is needed to improve their water quality to ensure breeding safety.

In Table 3.2, it can be seen the total nitrogen concentration of water in Dalian Lake water area is generally high, the total nitrogen content of water in its aquatic forest is 1.9 times of that of the Class V water quality index, so it is the poor Class V water; the total nitrogen in ponds is 2.435 mg/L averagely, which is 1.22 times of that of Class V water (according to the national standard), which is entrophicated seriously; the TN content in rivers in water source area is 2.42 mg/L averagely, which is entrophicated too. In view of the analysis on the investigation data, the water entrophication in water source area has a sequence as follows, *i.e.*: aquatic forest>pond>river. From which it can be seen the water that does not flow has high entrophication, which reveals the water source area is the pollution source too (Figs. 3.3 and 3.4).

Table 3.3 Bottom sediment depth of Dalian Lake

Site	Longitude and latitude	Water depth (m)	Offshore distance (m)	Sediment depth (cm)
T1	31°04.306; 121°00.611	1.15	3.0	11
T13	31°04.301; 121°00.491	1.75	4.5	18
T11	31°04.347; 121°00.415	0.75	1.0	8
River course 1	—	2.87	—	—
River course 2	31°04.177; 121°00.463	0.95	—	135
River course 3	31°04.283; 121°00.309	1.62	—	20
River course 4	31°04.361; 121°00.385	1.53	—	30
River course 5	31°04.593; 121°00.050	1.75	—	65
T71	31°04.331; 121°00.344	—	—	—
T63	31°04.535; 121°00.090	0.70	2.2	25
Aquatic forest	31°04.528; 121°00.101	1.2	1.5	40

The bottom material of the demonstration area at Dalian Lake is shown in Tables 3.3 and 3.4. Upon the comparison (open ground), the content of nitrogen, phosphor and organic matter is 0.24 g/kg, 0.52 g/kg and 0.96 % respectively, which reveals the nutritive salt in the bottom soil of Dalian Lake area is at a lower level. The aquatic forest has the roles and functions of filtration and pollutant purification, where the content of total nitrogen, total phosphor and organic matter is respectively higher than that of rivers and breeding ponds. The nutritive salt and the sediments in the rivers from which samples are taken come mainly from the domestic sewage from small towns and the village-level rivers, where the pollution is serious and the content of total nitrogen, total phosphor and organic matter is respectively 2.67 g/kg, 0.79 g/kg, and 3.82 %; each of which grows by 10.1 times, 0.53 times and 3.0 times of the comparison data. The nutritive salt and sediments in breeding ponds come mainly from the feedstuff and fertilizer, where the content of total nitrogen, total phosphor and organic matter grows respectively by 5.9 times, 0.47 times and 2.92 times of the comparison data.

3.3 Biological Characteristics of Water Environment

3.3.1 Density of Phytoplankton

During the investigation period in August 2008, the total average density of the phytoplankton in Dalian Lake water source was 2.04×10^7 cell/L; among which the maximum one was *Cyanophyta, i.e.*: 1.56×10^7 cell/L (accounting for 76.6 %); following in turns it was the *Chlorophyta* 3.57×10^6 cell/L (17.5 %), *Bacillariophyta* 7.22×10^5 cell/L (3.5 %), *Cryptophyta* 3.74×10^5 cell/L (1.8 %), *Xanthophyta*

Table 3.4 Analysis on bottom materials of Dalian Lake

	Analysis items		
Sampling site	Total nitrogen (g/kg)	Total phosphor (g/kg)	Organic matter (%)
CK (contrast)	0.24	0.52	0.96
Aquatic forest A	4.34	1.27	8.38
Aquatic forest B	1.77	0.93	4.05
Aquatic forest C	3.35	0.92	6.13
T1A	1.62	0.89	3.81
T1B	1.26	0.58	5.14
T1C	1.07	0.73	2.57
T11A	1.46	0.59	3.40
T11B	0.86	0.74	5.03
T11C	2.10	0.57	2.56
T13A	2.19	1.04	4.00
T13B	1.57	0.92	3.32
T13C	1.78	0.83	2.59
T63A	2.09	0.78	3.57
T63B	2.05	0.88	6.54
T63C	1.76	0.79	3.93
T71A	1.91	0.73	4.37
T71B	1.45	0.91	3.56
T71C	1.75	0.49	2.12
River course 2A	2.32	0.60	3.37
River course 2B	1.95	0.60	2.65
River course 2C	1.58	0.56	3.46
River course 3A	3.42	1.27	5.54
River course 3B	3.48	1.08	7.90
River course 3C	2.42	0.98	5.03
River course 4A	3.49	0.68	6.43
River course 4B	3.45	0.77	5.82
River course 4C	1.94	0.60	3.26
River course 5A	2.33	0.61	4.32
River course 5B	2.32	0.73	3.89
River course 5C	1.72	0.55	2.53

4.00×10^4 cell/L (0.2 %), *Euglenophyta* 3.92×10^4 cell/L (0.2 %) and *Pyrrophyta* 2.35×10^4 cell/L (0.1 %).

Among the Sampling Sites (Fig. 2.1), the density of the Sampling Site V was maximum, *i.e.*: 6.73×10^7 cell/L, following in turns it was the Sampling Site II (4.53×10^7 cell/L), Sampling Site X (3.12×10^7 cell/L), Sampling Site VIII (2.67×10^7 cell/L), Sampling Site III (1.85×10^7 cell/L), Sampling Site IX (8.75×10^6 cell/L), Sampling Site I (4.40×10^6 cell/L), Sampling Site VI (6.41×10^5 cell/L), Sampling Site IV (6.18×10^5 cell/L) and Sampling Site VII (2.90×10^5 cell/L).

Table 3.5 Average chlorophyll *a* content of phytoplankton in each sampling site

Sampling site	I	II	III	IV	V
Chlorophyll *a* content (μg/L)	9.11	136.73	92.93	7.63	50.43
Sampling site	VI	VII	VIII	IX	X
Chlorophyll *a* content (μg/L)	9.11	8.72	85.32	246.91	42.41

Table 3.6 Composition of density of zooplankton in Dalian Lake

Kinds	Average density (number/L)	Percentage (%)
Rotifera	7.1	44.9
Cladocera	3.7	23.4
Copepoda	5.0	31.7
Total average	5.3	100

Table 3.7 Density of zooplankton in every sampling site

Sampling site	I	II	III	IV	V
Density of zooplankton (number/L)	23.5	2.7	2.0	16.8	16.2
Sampling site	VI	VII	VIII	IX	X
Density of zooplankton (number/L)	3.1	2.9	1.1	22.1	67.4

3.3.2 *Chlorophyll* a *Content of Phytoplankton in Dalian Lake*

The average chlorophyll *a* content of phytoplankton in water in the investigation area was 68.93 μg/L, which fluctuated between 7.63 and 246.91 μg/L. The chlorophyll *a* content at the Sampling Site IV was minimal, but it was maximal at the Sampling IX; the chlorophyll *a* content in river courses was lower, but the one in ponds was higher (Table 3.5).

3.3.3 *Density of Zooplankton*

The average density of zooplankton in the investigation area in Dalian Lake water source area was 15.8 number/L; among which the density of the *rotifera* was maximal, *i.e.*: 7.1 number/L, which accounted for 44.9 % of the total density; the density of the *Copepoda* followed, it was 5.0 number/L, accounting for 31.7 %, and then it was the *Cladocera*, *i.e.*: 3.7 number/L, accounting for 23.4 % (Table 3.6).

The density of zooplankton in every Sampling Site was different from each other hugely, where the one at the Sampling Site VIII was minimal, *i.e.*: 1.1 number/L, but the one at the Sampling Site X was maximal, *i.e.*: 67.4 number/L. The density of zooplankton in the Sampling Site I, IV and V of the external river course and the Sampling Site X of the marsh wetland was higher a bit, but the one in every Sampling Site of ponds was relatively lower, expect the Sampling Site IX where the water was eutrophicated seriously (Table 3.7).

Table 3.8 Density of zoobenthos and biomass in Dalian Lake

Species	Mollusca	Oligochaeta	Chironomid larvae	Total average
Density (number/m^2)	614	2,655	603	3,872
Biomass (g/m^2)	472	10.3	5.8	488.1

3.3.4 Density of Zoobenthos and Biomass in Dalian Lake

Upon the investigation in 2008, it was found the total average density of zoobenthos in water in Dalian Lake water source area was 3,872 number/m^2, among which the one of oligochaeta was 2,655 number/m^2; following it was Mollusca (614 number/m^2) and Chironomid larvae (603 number/m^2). The total average biomass of zoobenthos in Dalian Lake was 488.1 g/m^2, where the one of the Mollusca was maximal, *i.e.*: 472 g/m^2; following it was the oligochaeta (10.3 g/m^2) and the Chironomid larvae (5.8 g/m^2) (Table 3.8).

Chapter 4
Analysis on Pollution Source

In recent almost 20 years, the Tai Lake watershed area has developed its economy rapidly and the township enterprises have been flourishing, but the wastewater treatment and governance level has been far behind than the increase of pollution discharge; plenty of wastewater were discharged into rivers, lakes and sea directly without treatment to cause all of them get serious pollution. Currently the industrial and domestic wastewater and sewage of 5,000,000,000 m^3 are directly discharged into rivers and lakes in the Tai Lake watershed every year. Among which, the percentage of the wastewater treated is only 20 % or less. According to the monitoring on the cross-section of main rivers and lakes in the Tai Lake watershed in 2000, including the main lakes, for instance, Tai Lake, Dianshan Lake; the main water supply courses in the watershed, for instance, Taipu River, Wangyu River; the main watercourses connecting the Tai Lake, for instance, the East and West Shao River, Nan River, Zhihu Port, Liangxi River, Xujiang River; and the provincial boundary watercourses, for instance, Hongqi Pond, Shanghai Pond, etc., the water quality at the monitoring cross-sections of 19.4 % can reach the surface water standard of Grade II or III (evaluation standard: GB3838—88); the water quality at the rest monitoring cross-sections of 80.6 % is found polluted seriously; among which the water quality at the monitoring cross-section of 48 % is Grade IV, 14 % Grade V, but 23 % is inferior to Grade V (such water has no any use value). The pollution type of the rivers in Tai Lake watershed is the organic pollution, and the nonconforming index of water quality is the permanganate index (COD_{Mn}), ammonia–nitrogen (NH_3-N), etc. Since plenty of nutritive salts (nitrogen, phosphor and other pollutants) go into lakes, the lakes are entrophicated seriously. Currently the Tai Lake, the maximum water supply source area in the watershed, is entrophicated seriously. In view of the monitoring data 2000, 71 % water area was entrophicated, and 29 % was in the middle of eutrophication.

S. An and L. Wang, *Wetland Restoration: Shanghai Dalian Lake Project*,
Springer Environmental Science and Engineering, DOI 10.1007/978-3-642-54230-5_4,

4.1 Internal Source Pollution

The internal source pollution in Dalian Lake area is caused by the aquatic breeding pollution and the water sediments. There are 336 mu ponds in the demonstration area, which are mainly used to breed grass craps, silver carps, bluntnose black breams, crucian carp and other fishes. According to the investigation, the acre yield of the breeding ponds in the water source area is 1,250–1500 kg/mu; the pond is fed the feedstuff by 2,880 kg/mu; where the feed protein content is 24 %, phosphor content is 1.65 %, and the feedstuff coefficient is 2.3.

According to the fish feedstuff nutriment transformation analysis (Fig. 4.1), it is needed to feed 867.68 t feedstuff in the breeding ponds in the demonstration area every year, which is equal to 37,160 kg nitrogen and 16,450 kg phosphor. Except 27 % nitrogen and 24 % phosphor convert into the water products, the rest go into water and go into the sediments completely.

The sediment pollution is the mostly important home of the nutritive salts of water environment. The nutritive salts are mainly from the solid particles in water, including the one from the degradation of animal and plant residues in water, exchange and absorption of nutritive salt, weathering and resolution of sediment, and the sedimentation of solid particles in water etc. The water sediment mass affects the water quality directly, so the evaluation on the water sediment mass is an important content of the evaluation on the quality of water environment.

Upon the comparison (open ground), the content of nitrogen (TN), phosphor (TP) and organic matter (TOC) is 0.24 g/kg, 0.52 g/kg and 0.96 % respectively, which reveals the nutritive salt in the bottom soil of Dalian Lake area is at a lower level. The aquatic forest has the roles and functions of filtration and pollutant purification, where the content of total nitrogen, total phosphor and organic matter is respectively higher than that of rivers and breeding ponds. The nutritive salt and the sediments in the rivers from which samples are taken come mainly from the domestic sewage from small towns and the village-level rivers, where the pollution

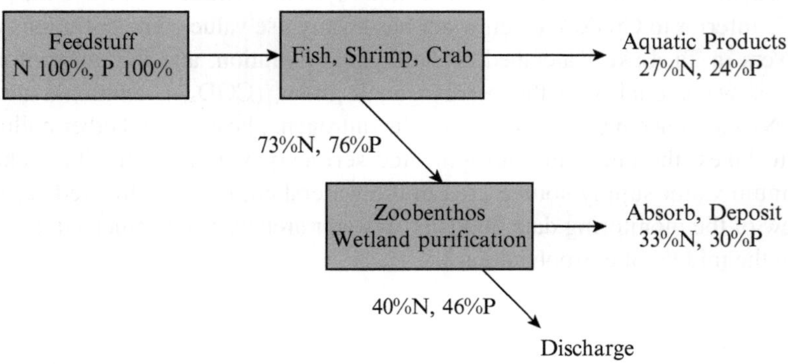

Fig. 4.1 Nitrogen-phosphor conversion relation of breeding ponds

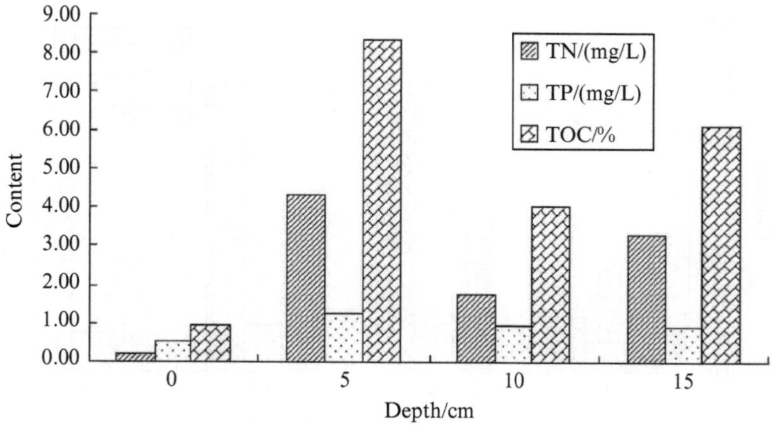

Fig. 4.2 TN, TP, TOC content in aquatic forest

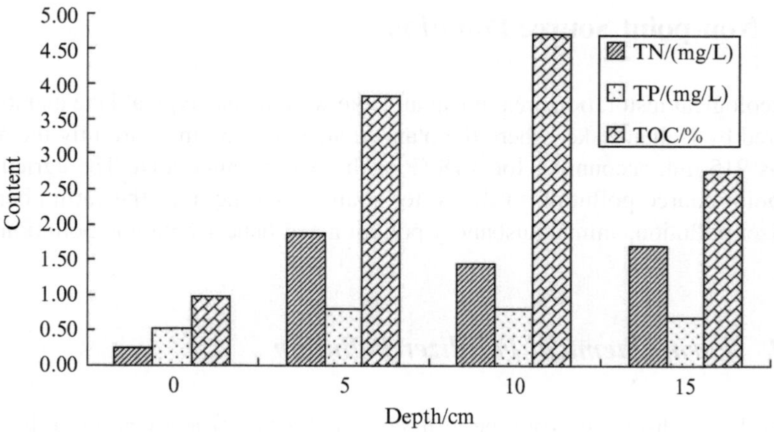

Fig. 4.3 TN, TP, TOC content in ponds

is serious and the content of total nitrogen, total phosphor and organic matter is respectively 2.67 g/kg, 0.79 g/kg, and 3.82 %; each of which grows by 10.1 times, 0.53 times and 3.0 times of the comparison data. The nutritive salt and sediments in breeding ponds come mainly from the feedstuff and fertilizer, where the content of total nitrogen, total phosphor and organic matter grows respectively by 5.9 times, 0.47 times and 2.92 times of the comparison data.

From which it can be seen a relation as follows, namely nitrogen sediment > organic matter sediment > phosphor sediment, whichever it is the rivers or the breeding ponds; which shows the nitrogen-contained organic substances are the main pollutants of Dalian Lake (Figs. 4.2, 4.3, and 4.4).

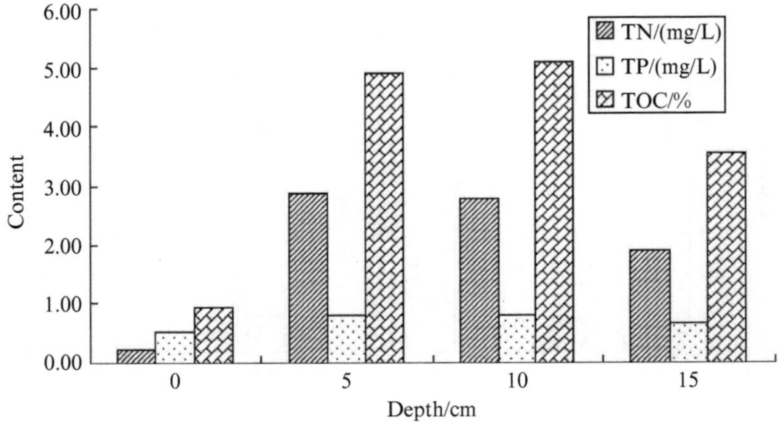

Fig. 4.4 TN, TP, TOC content in bottom materials of rivers

4.2 Non-point Source Pollution

The ecological restoration area in Dalian Lake wetland is a typical lake marsh zone centered by Dalian Lake, where the planned area is 14.6 km^2; currently the arable land is 915 mu, accounting for 4.18 % of the water source area. The agricultural non-point source pollution in the water source area includes the farm chemical fertilizer pollution, animal husbandry pollution and fishery breeding pollution.

4.2.1 Farm Chemical Fertilizer Pollution

The pollution discharge from the farm chemical fertilizer application in the water source area is COD 130.2 t/year, ammonia-nitrogen 30.9 t/year, and total phosphor 7.4 t/year. According to the investigation, many farmers, in order to get high yield and prevent and control diseases, use plenty of chemical fertilizers and farm chemicals, which cause the non-point source pollution to lands. According to the statistics, the non-point source pollution on farm land accounts for 22 % of the total non-point source pollution on the water source area.

4.2.2 Animal Husbandry Pollution

According to the investigation, the households in Dalian Lake water source area breed chickens, ducks, gooses and other poultries, although the large-scale poultry husbandry is prohibited, the household normally breeds 20–40 poultries; some households breed pigs, sheep and other large poultries. Due to the household breeds poultry in the area near to rivers or bays, the excrement of these poultries enters into the lakes and rivers as the rain flows, which causes the local water to be entrophicated.

4.2.3 Fishery Breeding Pollution

There are 7,046 mu fish ponds in Dalian Lake water source area, which accounts for 32.19 % of the water source area; additionally there are 1,700 m netted breeding areas in the lake area. According to the statistics, the pond in Dalian Lake area can discharge the COD 269.1 t/year, ammonia-nitrogen 4 t/year and total phosphor 1.7 t/year; the netted breeding areas can discharge COD 65.6 t/year, ammonia-nitrogen 1.4 t/year and total phosphor 6.4 t/year.

4.3 Point Source Pollution

4.3.1 Pollution by Rural Life

According to the investigation, it is found the pollution by rural life is mainly caused by the direct discharge of domestic sewage and random discard of house refuses and agricultural straws without treatment. The domestic sewage discharged by 2,650 households in Dalian Lake water source area is about COD 109 t/year, ammonia-nitrogen 10.9 t/year and total phosphor 7.1 t/year. The pollutant discharge coefficient of rural life is COD 40 g/(day-person) and ammonia-nitrogen 4 g/(day-person). The pollution caused by household is over one sixth of the total pollution, so the pollution is rather serious.

4.3.2 Pollution by Industries

In accordance with the investigation, there are totally 20 industrial enterprises in Dalian Lake water source area, including 6 enterprises under the Category I and 14 enterprises under the Category II.

The enterprises under the Category I are the pollution-free enterprises, which are: Shanghai Cenda Balance Co., Ltd, Shanghai Peninsula Gardening Co., Ltd, Lansheng Musical Instrument Co., Ltd, Shanghai Chenhe Trade Co., Ltd, Shanghai Zhijie Craftwork Weaving Co., Ltd and Shanghai Sanjiaodi Co., Ltd.

The enterprises under the Category II are the seriously pollution enterprises, which are: Shanghai Yiyan Garment Co., Ltd, Xicheng Bricks and Tiles Factory, Xinnong Feedstuff Co., Ltd, Qingpu Glass Instrument Plant, Liansheng United Timber Mill, Liansheng Water Pump Factory, Liansheng Coating Factory, Shanghai Yifa Electrical Appliances Co., Ltd, Shanghai Mingfu Braiding Factory and Yaye Carton Packaging Co., Ltd, etc.

The hazardous substances contained by the wastewater produced by these enterprises intensify the water pollution, for instance, the oil, benzene and copper contained by the wastewater discharged by the Qingpu Glass Instrument Plant, as well as the sulfide, nitryl-based substances, fibrin and washings contained by the wastewater discharged by textile mill.

Bibliography

Cheng, X., & Li, X. P. (2008). 20-year variations of nutrients (N and P) and their impacts on algal growth in Lake Dianshan, China [J]. *Journal of Lake Sciences, 20*(4), 409–419 (in Chinese).

Li, Y., Wang, L. Q., & Zhang, R. L. (2008). Nutrient release from decomposition of submerged macrophytes of Lake Dianshan [J]. *Environmental Pollution and Control, 30*(2), 45–49 (in Chinese).

Shen, G. X., Wang, Z. Q., & Qian, X. Y. (2010). The features of agricultural nonpoint source pollution in the Dalian Lake area of Shanghai [J]. *Acta Agriculturae Shanghai, 26*(1), 55–59 (in Chinese).

Sun, Z. Z., Wu, W. L., Lin, H. S., & Liu, S. M. (1998). The water quality monitoring on Dianshan Lake and upper reach of Huangpu River for fishery [J]. *Fisheries Science & Technology Information, 25*(5), 220–223 (in Chinese).

Wang, Z. Q., Shen, G. X., Qian, X. Y., & Zhu, Y. (2010). Seasonal impact of agricultural non-point source pollution on water environment of Dianshan Lake basin in Shanghai city [J]. *Agricultural Science and Technology, 11*(7), 83–86.

Part II
Project Planning, Design and Construction

Chapter 5
Wetland Restoration Strategy

5.1 Evaluation on Status of Dalian Lake Wetland

5.1.1 Analysis on Ecological Functions of Dalian Lake Wetland

The whole functions of the ecological system of Dalian Lake wetland include the hydrologic regulation function, water quality purification function, substance production function and wetland tourism, etc. The main mechanism of the wetland system to control the non-point source pollution is to increase the runoff infiltration, delay runoff flow, increase dwelling time, etc., and then make pollutants detain in the system to decompose and transform, and finally reduce the pollutant loads from going into the downstream water body; meanwhile, the rainfall runoff intercepted shall be recycled, which can improve the utilization ratio of water sources and relieve the scarce water source. When designing the wetland system for controlling the non-point source pollution, it is needed to consider the characteristics of rainfall runoff (rainfall strength, rainfall duration, rainfall time interval, etc.), wetland dewatering time, drought tolerance of plants, transpiration and other factors so as to make the structure and functions of wetland fit to the randomness and rapid change of rainfall. The load of unit area affects the degradation efficiency of the wetland on pollutants; the improper design can make wetland system unworkable; in other words, the primary issue of the design work is to determine the pollution load of wetland area (currently it is needed to determine the pollutant reduction objective of wetland in accordance with the pollution source investigation result, and then obtain the pollution load of wetland area). In a wetland, the degradation extent of pollutant is related to its dwelling time; whereas the dwelling time is inversely proportional to the unit area's load of wetland. The non-point source pollutant, retention rate of runoff and runoff process, dwelling time, water depth, etc. have close relation; however, the dwelling time of the runoff in the wetland system is controlled by the runoff inflow process.

S. An and L. Wang, *Wetland Restoration: Shanghai Dalian Lake Project*,
Springer Environmental Science and Engineering, DOI 10.1007/978-3-642-54230-5_5,

5.1.1.1 Hydrologic Regulation Function of Dalian Lake Wetland

The wetland is the transition belt between open water body and land, with particular hydrologic characteristics. The wetland hydrology includes mainly the water intake, output, water depth, water flow mode, water-flooding duration and water flooding frequency. The water input source includes rainfall, ground water and overflew water; the water output channel includes water evaporation, surface water flow and infiltration.

The hydrologic regulation situation of wetland, *i.e.*: the water cycle, varies with the four seasons, which is mainly subject to the water balance, landform, soil, geology and underground water's characteristics of the wetland. For the general equation of Liu Houtian, the tidal intake and output is removed; where the formula is expressed as follows, *i.e.*:

$$\Delta V = P_n + S_i + G_i - E_t - S_o - G_o$$

Where V is the impounding water volume of wetland; ΔV is the wetland's impounding water volume change; P_n is the effective precipitation; S_i is the surface water inflow, including the overflew water of rivers; G_i is the underground water inflow; E_t is evaporation; S_o is the surface water outflow; G_o is the infiltration.

The hydrologic regulation function of the wetland is to regulate the impounding water, flood, runoff, surface water and underground water through water impounding, drainage and evaporation. The water circulation regulation capacity of the surface water in Dalian Lake wetland can be calculated by the following formula, *i.e.*: $V = V_{lake} + V_{pond} + V_{rice} + V_{depression} = 4{,}648{,}300 + 3{,}053{,}000 + 610{,}000 + 1{,}000{,}000 = 9{,}311{,}300\ m^3$ (about 9.31 million m^3) (this is also the impounding capacity of Dalian Lake).

5.1.1.2 Water Quality Purification Function of Dalian Lake Wetland

The wetland ecological system has the water source purification function; it is called as the "kidney" of the nature. The wetland absorbs and discharges nutriments and other chemicals upon the hydrologic ways (rainfall, surface water runoff, groundwater flow), where a series of biological localization processes occur in the particular oxygen-deficit reducing condition and the environment of the thick deadwood and dead leave coverage of the wetland ecological system, for instance, the oxidization process and reduction process of nitrogen fixation, denitrification, photosynthetic carbon fixation and marsh gas treatment, phosphor, sulphur and other elements, as well as the absorption and release process of mineral substances, etc. Such water source purification function of the wetland ecological system plays an important role in balancing the substance circulation of the wetland system, even the whole world. However, the biochemical process dominated by the substance balance principle is rightly the best natural tool to improve water quality. As per the analysis on the water environmental capacity of Dalian Lake, see the following.

Calculation of Water Environmental Capacity

The water environmental capacity is the maximal allowable pollution load or pollution-receiving ability required by the water environment quality standard; the water environmental capacity is based on the environmental objective and the self-purification law of water.

1. Calculating factor: COD, TN, TP.
2. Object: Dalian Lake wetland.
3. The desired value of water quality is subject to the national standard for environmental quality of surface water (GB3838—2002) (III Standard), *i.e.*: COD is 20 mg/L, TP is 0.05 mg/L and TN is 1.0 mg/L.

Mathematical Model for Calculation of Water Environmental Capacity

1. COD.

$$W_c = C\left(3.65K_cV + 31.536Q_{out}\right) - 31.5636Q_{in}C_o$$

Where, W_c: COD Environmental capacity (kg/day); C: desired value of water quality (mg/L); C_o: background concentration of water quality (mg/L); K_c: COD degradation coefficient (1/day); Q_{out}: Lake's outflow (m^3/s); Q_{in}: flow of incoming water under the design conditions (m^3/s).
2. Total phosphor.

$$I_p = P(\alpha A + 31.536Q_{out})$$
$$\alpha = 11.6 + 0.2H\mathbf{e}$$
$$\varepsilon = 3153.6Q_{out} / V$$

Where, I_p: total phosphor's environmental capacity of lakes (kg/day); P: total phosphor water quality's desired value of lakes (mg/L); α: sedimentation coefficient of phosphor (m/a); A: area of lake (km^2); H: depth of lake (m); ε: erosion coefficient of lakes (l/a); V: impounding capacity of lakes (10^4 m^3). The meanings of the other symbols are as same as the ones above.
3. Total nitrogen.

$$I_n = N\left(\alpha A + 31.536Q_{out}\right)$$

Where, I_n: total nitrogen's environmental capacity of reservoir (kg/day); N: total nitrogen water quality's desired value of reservoir (mg/L). The meanings of the other symbols are as same as the ones above.

Table 5.1 Hydrologic parameters of Dalian Lake

Year	Real flow (10^4 m^3)	Water intake (10^7 m^3)	Intake flow (m^3/s)	Water outflow (10^7 m^3)	Outgoing water flow (m^3/s)	Water surface area (km^2)	Water depth (m)	ε	α
High flow year	1,740	17.0	165.7	50	2.41	8.7	2.0	0.1875	11.6
Normal flow year	1,357.5	13.0	27.2	30	1.70	8.7	1.8	0	11.6
Low flow year	975	9.0	2.51	20	0.99	6.5	1.5	0	11.6

Definition of Hydrologic Parameters

Dalian Lake inflows and outflows continuously. Hereinafter the water environmental capacity of Dalian Lake in different normal flow years shall be calculated; where the hydrologic parameters involved are shown in Table 5.1.

The COD background concentration is 10 mg/L. Dalian Lake flows slowly (pollutants degrade slowly), however, due to the existence of aquatic organisms and the aeration of lake surface, pollutants can be degraded. By referring to the other research achievements, the limit of 0.05/day is adopted as the COD degradation coefficient (K_c).

Calculation Result

The environmental capacity of the total phosphor and the total nitrogen of Dalian Lake in the normal flow year is 1.8 t/a and 22.5 t/a respectively; and its COD environmental capacity is 430.7 t/a.

5.1.1.3 Substance Production Function of Dalian Lake

The energy of phytoplankton and aquatic bacteria is mainly used by four ways; firstly, it is ate directly by silver carp, Parabramis pekinensis and other filter feeders; secondly, it is ate by zooplankton and then by filter feeders; thirdly, it is ate by Zoobenthos and then by trash fishes; fourthly it is ate by trash fishes and then by other fishes. The Zooplankton is not only ate by Aristichthys nobilis, but also by mild predacious fishes and wild fishes (Fig. 5.1).

5.1.2 Integrated Evaluation on Status Quo of Dalian Lake Wetland

For the ecological system of Dalian Lake, its food web structure is not only simple, but also its ecological system is weak, it is intervened frequently by people and some links of them are blocked, for instance, there are large quantities

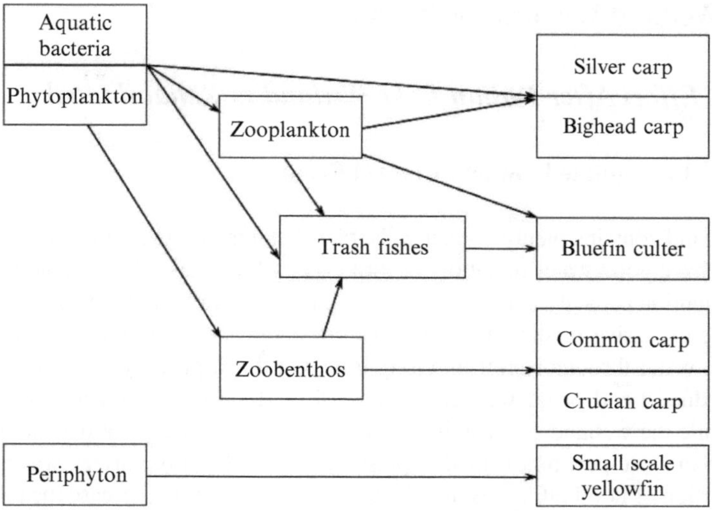

Fig. 5.1 Main flow and utilization ways of energy in Dalian Lake

of aquatic algae, high primary productivity, and the high-end fish resource structure is not perfect and so on.

The water and the bottom material of Dalian Lake is polluted seriously, even the water is eutrophicated; the pollution source comes mainly from the self-breeding, plantation, living, external water inflow, etc. Meanwhile the water conservation and the loss of the water purification function can also form the pollution source. The natural landform and structure of Dalian Lake is messed and the water system is not open, its wetland's landscape function is not at a high level.

The local economy is backward, industrial structure is simple and resident's income is low. Among the first 625 mu demonstration area of Dalian Lake water source area, the fishing ponds account for 59.85 %; so the economic output is the fish. According to the investigation, the yield of the fishing pond is 1,250–1,500 kg/mu, and the fishery output value is RMB 5,100,000/year–6,100,000/year; there are 8 breeding households, the average output value of every household is RMB 637,500/year–762,500/year; by deducting the cost factors, average income of every household is around RMB 50,000/year. Such income level is slightly higher than that of the local residents, but is far from the national modern index.

For the ecological restoration area of Dalian Lake water source area, where the land in the demonstration area has already been transferred to Dianshan Lake Co., Ltd through the way of land transfer, and some residents work in Dianshan Lake Co., Ltd; meanwhile the local residents receive the land transfer use fee from Dianshan Lake Co., Ltd. According to the investigation, the local rural residents are satisfied with the land use in such way; which creates conditions for implementing the ecological restoration works in Dianshan Lake water source area.

5.2 Wetland Restoration Thinking

5.2.1 Effect After Dalian Lake Wetland Is Rehabilitated

5.2.1.1 To Regulate Flow and Control Flood

The wetland contains plenty of peat soils, plants and massy impermeable stratum; all of them have good water retention performance; which make the wetland have huge water retention capacity and it is equal to a huge impounding reservoir. The wetland can store excessive precipitation in rainstorm period and flooding period, and then drain the water through runoff in a long time. On this aspect, the wetland can effectively reduce flood peak, weaken the hazard of flood to the downstream reaches. Meanwhile the wetland can also impact the local hydrologic situations and impound flood, where one portion of flood is preserved in the form of surface water, and the other portion of flood infiltrates into the underground water to increase the reserves of underground water; such underground water can relieve the scarce water resource in the water scarcity season and then the purpose to regulate the water source is achieved.

5.2.1.2 To Purify and Filter Water Quality

The wetland is helpful to relieve water flow. When the flood containing pollutants flow through the wetland, it may slow down, which shall be beneficial to deposit and expel toxics and foreign matters. Some wetland plants, just like *Phragmites communis* Trin, Eichhornia crassipes, etc., can effectively absorb toxic substances and play the role of purification and filtration, and then protect the water quality of the downstream reaches. Meanwhile the wetland plants can also remove the nitrogen, phosphor and other nutritive components in water, for instance, the average removal rate of *Phragmites communis* Trin on TN and TP can reach 74 % and 97 % respectively; whereas the average removal rate of *Oenanthe javanica* (Blume) DC on TN and TP can reach 87.22 % and 94.77 % respectively; the average removal rate of *Vetiveria zizanioides* L. on TN and TP can reach 77.94 % and 82.86 % respectively. All wetland plants said above can efficiently remove the TN and TP in water. Whilst water flows through the wetland, the nutritive components in water can be absorbed by the wetland plants or deposited into the wetland sediments; such process can also purify the water source quality for the downstream reaches. Additionally, except the wetland substrate works as the fill framework or provides the attachment surface to microorganisms, the process that the wetland substrate absorbs or exchanges ions can also produce the efficacy of pollutant purification.

5.2.1.3 Efficacy of Air Purification

The wetlands ecosystem can impact the composition of atmosphere, absorb carbon dioxide and release plenty of oxygen. Some researches show the small particles in air, for instance, dust and bacterium, fungus and other microorganisms, are inclined to move towards the wetland; therefore the wetland can produce the efficacy of purifying the air in the surrounding areas.

5.2.1.4 Impact of the Wetland on Microclimate Environment

The observed results reveal the evaporation of the wetland is 2–3 times of that of the water surface. An intensive evaporation can increase the air humidity, reduce the temperature of surrounding areas; meanwhile after water evaporates and forms steam, it goes to the surrounding area in the form of precipitation to preserve the local humidity and precipitation. This is the cold-humid effect of the wetland. When the temperature is high and the evaporation increases, the cold-humid effect of wetland shall become more obvious.

The moist of wetland is richer than that of the surrounding land; however, when water has higher calorific capacity, its temperature shall change slowly; therefore the wetland has the function of delaying air temperature change, especially when air temperature rises. In view of the general tendency that the global climate warms up, this point is very important, *i.e.*: the wetland can maintain the air temperature of the area stable relatively; meanwhile the overall amplitude of variation of the ground temperature of wetland is also less than that of the surrounding area, which can offer a relatively stable ground temperature environment to animals and plants.

5.2.2 Restoration Strategy of Dalian Lake Wetland

5.2.2.1 Adjust Land Use Form Rationally, Conduct Large-Scale Production and Control the Non-point Source Pollution to Agriculture

Adopt the land transfer mechanism to transfer the lands and ponds in Dalian Lake area; adopt the company system or cooperative operation to plan and use land; make the agricultural development planning, control agricultural non-point source pollution, improve land yield and walk on a road of sustainable development according to the local characteristics.

5.2.2.2 Change the Existing Agricultural Production Mode, Adjust the Industrial Structure and Carry Out the Organic Agriculture Mode

The ecological fishery is a fishery pattern to utilize, protect and cultivate fishes by using the ecological principle, systematic and scientific method, absorbing the technical essence of the traditional fishery and using the modern scientific techniques, which is featured in ecological ration, coordinate function, resource regeneration, economic efficiency, virtuous cycle and intensive farming and management. Such ecological fishery focuses not only on economic benefit, but also ecological effect and social effect. The ecological fishery is the only road to achieve the sustainable development of fishery. The theoretic systems to support the ecological fishery include the ecological balance theory, species mutualism theory, food chain theory, ecological niche theory, multilevel and hierarchical use theory, dissipative structure theory, hierarchical system theory, homeostatic mechanism theory, fringe effect theory and ecological economics theory, etc.

In accordance with the characteristics of fishery resources and environment of Dalian Lake, the preferred ecological countermeasure is to breed silver carps and big head carps to control and use the plankton and the aquatic bacteria; protect the breeding of Xenocypris microlepis so as to control and use the periphyton; protect the common carps, crucian carps and other omnivorous fishes so as to utilize the large zoobenthos; propagate the small ferocious fishes to control the trash fishes; avoid the error arising from introducing species repeatedly and indiscretion and prevent hazardous external fishes from invasion.

Achieve the purpose to control planktonophage to get balance and control the aquatic ecosystem by breeding and propagating important fish (native) population structure. The investigation result shows there is no obvious correlation between the existing quantity of the phytoplankton and that of the zooplankter in Dalian Lake water area. On the other hand, the reduction of the filter fishes in this water area results in the increase of the phytoplankton. The combined action of the filter fishes and the zooplankton is a path to control the phytoplankton; where the filter fishes, by eating the phytoplankton, cannot only shorten the food chain, but also reduce the energy loss; moreover the nutritive substances in water, after such fishes are caught, can be removed in a great quantity; so it can prevent the water from getting entrophication.

The ecological fishery in Dalian Lake can focus on the lake-oriented ecological fishery—ecological maintenance, and then develop the recreational fishery appropriately to drive the local ecological agriculture and its by-line to develop.

5.2.2.3 Adjust the Living Quarter Layout of Residents

In accordance with the overall planning scheme of ring Dianshan Lake, the existing 47 natural villages shall be combined into 32 central villages in 2010; this plan shall gradually help to convert the existing rural population into the urban population, reduce the quantity of central village, and then achieve the development objective of "3-centralization".

5.2.2.4 Promote Ecological Courtyard Construction, Dispose House Refuses and Agricultural Wastes, and Change the Living Style of Residents

Construct the independent village/township sewage treatment system in the central village, treat and dispose sewage and wastewater in the "dispersed" mode or the "centralized" mode.

Popularize the ecological courtyard in countryside, popularize the solar energy, marsh and other energies, try to popularize the economic and ecological courtyard mode, and improve the living quality and the environmental consciousness of residents.

5.2.2.5 Design Rational Layout for Water Source Area

By taking advantage of the beneficial conditions of the integrated governance on Dianshan Lake, design and plan the functional layout of Dalian Lake water source area, make efforts to form the water source conservation purification zone combined with aquatic forest, field-swinging ecological fishery, economic aquatic plant and others, and tap the natural landscape value of the water source area sufficiently.

5.2.2.6 Build the Landscape Relief, Dredge the Water System Structure of Water Source Area

By taking advantage of the existing landform and geomorphological structure of the water source area, design and plan the water system structure of the demonstration area to let the water of Lianhu River and Douji Port flow into the water source conservation purification zone to exert the integrated wetland functions of Dalian Lake.

5.2.3 Management Measures of Wetland

Since Dalian Lake area is one of the core regions of Shanghai water area, the policies of many water source areas cause certain impact on the local economy and life. On one hand, the water source protection policy make the local residents hard to develop economy as same as the residents in other areas, which causes the local backward economic situation; on the other hand, the corresponding compensation measures and the development measures lag behind, which make the local residents poorer. Such vicious cycle causes an unbalanced economic, environmental and social development to the water source area.

Upon the investigations, it is found there are multiple problems in the water source area, *i.e.*:(1) lands are under the dispersed operation mode; it is hard to conduct the large-scale production and control the non-point source pollution; (2) extensive fishery production, the breed structure is irrational; the farm cultivation focuses on the traditional farming mode, the non-point source pollution of agriculture is still serious; (3) residents are not sensitive to environmental protection, where the local healthy facilities and living habits are backward and the living pollution is serious; (4) the traditional agricultural products are at a low benefit level, farmers are not interested in agriculture too much.

In view of the problems above and the investigations on the project, it is proposed to take the following measures for sustainable management, *i.e.*:

1. Let the local government to lead to make objectives and establish the true administration company or agricultural cooperation organization for Dalian Lake Water Source Area, attract the social and the non-governmental organizations and the local residents to participate in production and management, and make the management standards for sustainable development.
2. According to the regional industrial characteristics and the post-governance environmental advantages, implement the brand strategy and carry out the corresponding technical management mode.
3. Let the environment impact the behaviors, change the living behaviors of residents by popularizing the ecological courtyard construction.
4. Develop economy and improve the living level of residents, restrain residents' production behaviors through market mechanism, and promote Dalian Lake environmental protection and economy to develop by changing the production behaviors of residents.

5.3 Philosophy and Principle for Planning and Designing on Wetland Restoration Project

5.3.1 Design Philosophy of Wetland Restoration

In view of the geographical location of the project area and the significance of water quality safety guarantee in Shanghai Municipality, as well as the objective of the demonstration project, namely "construct harmonious community and re-build a city in regions south of the Yangtze River", the project shall focus on the philosophies as follows, *i.e.*:

1. Water source area protection philosophy-adhere to the restoration ecology principle to rejuvenate and re-build lake wetland and enhance the purification function of wetland and water.
2. Wetland restoration philosophy-rejuvenate the seriously entrophicated fishing ponds into the natural wetland and construct them into the wetland restoration model in the ring Tai Lake area, and then rebuild a city in regions south of the Yangtze River.
3. Philosophy to combine the economic and wetland ecological benefit-whilst rejuvenating wetland ecology, improving ecological benefit and service function, consider sufficiently the importance of economic benefit to make the wetland get certain economic output, and increase the income of the local residents through appropriate administration.
4. Harmonious community philosophy-construct a demonstration village in Huangpu River water source area, combine the wetland ecology restoration project and

new socialist countryside construction, convert the agricultural production mode, implement the wetland ecological rehabilitation project, and construct (improve) new modes of the new socialist countryside.

5.3.2 Design Principle of Wetland Restoration

Under the guide of the aforesaid restoration and rehabilitation philosophies, the project shall be constructed by following the principles as follows, *i.e.*:

1. Natural principle.

   Due to the intensive interference of human activities to the ecosystem of Dalian Lake, the banks, deposited sediments and water quality, as well as the natural geomorphological situation and nature of the local wetland have been hugely changed; therefore it is impossible to rejuvenate or restore to its former original condition before getting the interference of human activities. However, the wetland restoration should still make the lake ecosystem return to nature. Whilst making design, it is preferred to conduct the nature-oriented reconstruction so as to improve the natural environment and landscape, restore the functional characteristics of wetland and promote the virtuous circle of lake ecosystem.
2. Ecological integrity principle.

   An ecosystem is the product of the long-term evolution and development. All parts of an ecosystem, through their collaborative evolution, become its integral parts. For the wetland protection research, the ecosystem is the combined ecological system integrating with "society", "nature" and "economy"; such ecosystem is a networked system under the correlation, mutual complementation, interaction and reciprocal causation of the physical and chemical compositions of varied surviving organic matters and non-living things. Therefore the wetland protection and restoration should be focused on the ecological system level, ecological system structure and functional coordination.
3. "In line with local conditions" principle.

   Whilst implementing the wetland restoration and protection, it is preferred to encircle the local natural, social and economic conditions so as to guarantee the wetland ecosystem to run rationally. Take advantage of the existing landform, topographic features and conditions in the demonstration area, follow up the local conditions to make engineering design; meanwhile, follow the principle of biology and landscape aesthetics to design rationally to reduce engineering quantity to protect nature, economy and environment.
4. "Protect biodiversity" principle.

   The biodiversity of the fortune of the nature, but also the essence of the nature for continuous reproduction breed in an endless succession. The protection on biodiversity is not to protect the diversity of species, but protect the diversity of ecosystem and genetic resources. In the protection and restoration process of the flora and fauna resources of wetland, anyone should not only consider the quantity of one or multiple animals and plants, but the whole biologic chain; additionally it is required to advocate to rejuvenate the wetland habitats, and improve the population quantity by improving the wetland habitats.

5. Demonstrative principle.

   The project is aimed to become a wetland restoration mode with the demonstrative value, therefore the project design should focus on technical innovation, solving all engineering difficulties, control engineering cost, considering engineering security, operability and landscape purification efficacy, and then finally provide a popularization-oriented model for the Dianshan Lake wetland restoration project (in the west suburb of Shanghai) and the whole Tai Lake watershed for integrated water environmental governance.

6. Economic principle.

   This project is not a simple ecological landscape project, but a new wetland project capable of producing economic output and being utilized by local residents. Therefore the project design should sufficiently consider the economy of the wetland. In the project design, the aquatic vegetables and fishes beneficial for water purification and healthy water ecosystem and available to produce ecological effect and economic value are the binding point of the economy and ecology.

5.4 Overall Planning for Wetland Restoration Project

5.4.1 Project and Its Significance

Under the support of the "HSBC Climate partnership" Program (hereinafter referred to as "HSBC II"), the World Wild Fund for Nature Shanghai Office, based on its concern on the drinking water safety of Shanghai superlarge city and the demands on solving the agricultural non-point source pollution of its water source area and the community life pollution, made multiple field investigations and repeated integrated evaluations, and then finally chose Dalian Lake community in Dianshan Lake area in the upper reach of Huangpu River as the demonstrative project initiation zone.

The Dalian Lake wetland restoration demonstrative project has the following significances, *i.e.*:

1. It is the need to guarantee the drinking water safety of Shanghai superlarge city.
2. It is the need to coordinate urban development and ecological guarantee of Shanghai city.
3. It is the need to make integrated governance on the water environment of ring Tai Lake watershed.
4. It is an important measure of Shanghai Expo. To create "ecological Expo."
5. It is the need to carry out "*General Planning of Shanghai for Qingxi area* 2005–2020".

5.4.2 Project Planning Objective and Overall Thinking

The planning objective of Dalian Lake wetland restoration project has the following main objectives, *i.e.*:

1. Establish the community-participation type wetland restoration and management mechanism, guarantee the long-acting operation of the wetland restoration project.
2. Establish the organic agricultural mode of wetland and offer organic products to local residents.
3. Re-establish wetland habitant diversity and improve the wetland biodiversity.
4. Rejuvenate wetland vegetation and re-construct healthy aquatic ecosystem.
5. Exert the wetland function of "kidney" of global to process wastewater and improve water quality of water source area.

The main thinking for the overall planning of the project is to distinguish the main pressure of Shanghai non-point source pollution, explore community-participation type water source area protection mode, rebuild Lianhu village and rebuild the urban water source through three important measures (healthy wetland, organic agriculture and partnership action), and then form the action principle with PEAR WETLAND (Pilot Ecological Administration and Restoration of Wetlands).

1. Healthy wetland: rejuvenate and rebuild wetland, construct water system food chain and establish healthy aquatic ecosystem through "fishing pond returning to wetland".
2. Organic agriculture: develop wetland organic agriculture, create "zero emission" agricultural mode, establish rural wastewater processing system and reduce domestic sewage pollution load.
3. Partnership action: conduct "1 + 1" partnership action, improve the local people's sense of environmental protection, increase residents' income and make local residents participate in the water source area protection voluntarily.

5.4.3 Planning Content and Scale of Project

The total planning area of Dalian Lake wetland restoration project is about 2,000 mu (130 ha), covering 6 subareas and 2 special projects, *i.e.*: divide the demonstrative project area into 6 subareas, namely wetland restoration and reconstruction area (Subarea A), forest wetland cultivation area (Subarea B), shoal wetland restoration and biodiversity conservation area (Subarea C), wetland organic agricultural area (Subarea D), coastal wetland restoration area (Subarea E) and community domestic sewage treatment area (Subarea F) and subordinating rivers' water system rehabilitation project (G) and effect monitoring project (H).

Wetland restoration and reconstruction area (Subarea A). Its planned area is 30 ha, the planning content includes: to relieve internal source pollution, recover wetland functions, strengthen wetland's purification capacity; increase the wetland biospecies diversity, establish healthy aquatic ecological system, coordinate to develop wetland vegetation and aquatic vegetable, and create the wetland organic agricultural mode.

Forest wetland cultivation area (Subarea B). Its planned area is 25 ha, the planning content includes: to cultivate forest wetland, re-build Tai Lake watershed forest wetland type, increase water fluidity, improve wetland purification capacity, richen wetland spatial hierarchy and provide habitats for birds.

Shoal wetland restoration and biodiversity conservation area (Subarea C). Its planned area is 10 ha; the planning content includes: to recover the wetland type dominated by shoal to provide feeding area for wading birds, reconstruct deep-water aquatic ecosystem to provide food source for swimming birds, richen plant and fish species, increase birds diversity.

Wetland organic agricultural area (Subarea D). Its planned area is 45 ha; its planning content is to reconstruct organic fishing pond: wetland purification system, develop "zero emission" organic fishing mode, create environmental friendly agriculture and farming mode, reduce agricultural non-point source pollution.

Coastal wetland restoration area (Subarea E). Its planned area is 5 ha; its planning content is to rejuvenate and reconstruct Dalian Lake coastal wetland and improve the self-purifying capacity of lake.

Community domestic sewage treatment area (Subarea F). Its planned area is 15 ha, which covers about 30 farmer households; its planning content is to construct rural domestic sewage treatment system to make water quality meet the national drainage standard. Change the non-environmental friendly behaviors through propaganda, education and training, improve residents' sense of participating in water source area protection and capacity.

Rivers' water system rehabilitation project (G): to reconstruct water system in project zone, enhance water fluidity, rejuvenate river's ecosystem and activate rivers' self-purifying capacity.

Effect monitoring project (H): to set the water quality monitoring point, make follow-up monitoring and evaluation on wetland purification effect and treatment efficiency of rural wastewater; set birds, fishes and plants monitoring points (stations) to make long-term monitoring on biodiversity; arrange the long-term water quality monitoring equipment to monitor dynamically the non-point source pollution of agriculture and fishery.

5.4.4 Anticipated Effect of Project

The anticipated effect of Dalian Lake wetland restoration project includes five aspects, namely environment, economy, society, ecology and landscape.

5.4.4.1 Environment

The rural sewage shall reach the national Class A drainage standard after being treated; the organic farmland and the organic fishing ponds shall be advocated the "zero pollutant discharge"; the sewage and wastewater passing through the treatment zone shall be purified from Class V to Class III.

The main technical indexes include:

1. Rural domestic sewage treatment: daily treatment sewage 30–40 t, operating cost RMB0.15/t, RMB4.5/day ~ RMB6.0/day; COD reduction 83–96 %; ammonia nitrogen reduction 81–98 %, total nitrogen reduction 71–97 %, total phosphor reduction 65–67 %; reach the national sewage and wastewater treatment drainage standard, *i.e.*: Class A water quality.
2. Wetland treatment system (wetland restoration & reconstruction area, forest wetland cultivation area, shoal wetland restoration and biodiversity conservation area): the healthy wetland can remove COD 60–80 %, annual removal of total nitrogen 3,650 kg (10 kg/day), total phosphor 730 kg (2 kg/day); daily treatment of poor Class V water (Dalian Lake area COD 15.1 mg/L, TN 4.88 mg/L, TP 1.05 mg/L) 2,350–2,580 t (m^3); water's suspension duration 5 days; water load about 10–15 cm/m^2; reach the national Class III standard for surface water (COD 6.0 mg/L, TN 1.0 mg/L and TP 0.2 mg/L).
3. Organic agricultural mode: to realize "zero pollution discharge", make drainage water quality meet national Class III standard for surface water.

5.4.4.2 About Economy

The wetland restoration and reconstruction area shall provide the extra economic income of RMB 120,000/year roughly to local residents every year; the organic fishery and the organic agriculture shall make the community economic income increase over 50 %; meanwhile the local residents' employment opportunity shall be increased.

The main technical indexes include:

1. Aquatic economic plants: plant *peperomia tetraphylla*, *Oenanthe javanica* (Blume) DC, *Sagittaria sagittifolia* L., *Eleocharis dulcis*, *Ipomoea aquatica Forsk*, lotus root and other economic plants in the wetland restoration and reconstruction area by 7.5 mu at least; the six plants aforesaid shall provide the economic income by RMB 40,000 or more.
2. Aquatic economic animals: fishes: 30 kg/(mu × year) · 150 mu = 4,500 kg/year; 4,500 kg × RMB12/kg = RMB 54,000; shrimp, shell and crab: 15 kg/(mu × year) · 150 mu = 2,250 kg/year; 2,250 kg × RMB 20/kg = RMB 45,000; income: RMB 99,000; young animals' cost is about RMB 15,000; the total net income is about RMB 84,000.

3. Organic agricultural mode: the yield shall be reduced by 30 % roughly, but the price shall be improved by 100–200 %; so the benefit shall increase by 40–110 %; additionally the farming cost shall be reduced by 50 % roughly, so the total income shall be improved by 50 % or more.

5.4.4.3 About the Society

To improve residents' sense of ecological environmental protection, improve residents' voluntariness for water source area protection, create the community-participation type water source area protection mode, establish the PEAR Wetland mode by integrating the demonstrative project; popularize the experience of the demonstrative project through media, thesis papers and books.

The main technical indexes include:

1. To provide Shanghai the water meeting Class III surface water quality standard by 868,700–956,300 t/year; among which the sum of the total qualified water from the rural domestic sewage treatment and the total qualified water from the wetland (through its purification function) is 10,950–14,600 t; the qualified water from the wetland treatment system is 857,750–941,700 t.
2. Whilst implementing, keeping and operating the demonstrative project, the local residents shall be absorbed to participate and which shall create 4–10 employment positions.
3. There are 800 people in Lianhu Village, 20,000 people in Jinze Town, Liantang Town and Zhujiajiao Town, and 200,000 people in Qingpu; all of them shall be educated to have the sense of water source protection; meanwhile the 50,000 habitants in water source area community shall change their living styles to participate in water source area protection.
4. Whilst constructing, keeping and popularizing the demonstrative project, 20 cooperative partners shall be developed; meanwhile it is planned to post 50 reports in local media, 20 reports in Shanghai media and 10 reports in national media; all people and organizations in the world are welcome to visit.
5. Improve the residents' sense of ecological environmental protection, build new socialist countryside, provide demonstration for wetland restoration and water source area protection in Qingpu, Shanghai, Tai Lake watershed and the areas in the lower middle reaches of the Yangtze River.

5.4.4.4 About Ecology

Increase biodiversity (increase 40 species or more), rejuvenate healthy aquatic ecological system, rejuvenate wetland (rivers, marshes, lakes) purification function, strengthen wetland treatment capacity to cope with the global climate change; reduce the carbon dioxide emission by 130 t/year approximately.

The main technical indexes include:

1. Wetland restoration: increase 40 species at least, increase biodiversity, create the healthy aquatic ecological system comprising 4 nutritive hierarchies at least.

2. Appreciation of ecological fixed assets: For the traditional high-yield fishing ponds and wetland in the demonstrative project area, the waste treatment, biodiversity protection and recreational cultural value of the former is "zero"; however, the latter can produce the ecological service values of RMB 16,050/($hm^2 \cdot$year), RMB 2,210/($hm^2 \cdot$year) and RMB 4,910/($hm^2 \cdot$year); the demonstrative project of 625 mu (41.7 hm^2) can provide the ecological service value of RMB 966,190/year approximately.
3. Carbon dioxide emission reduction: the wastewater treatment plant shall emit the carbon dioxide by 0.30 kg approximately when treating 1 t wastewater (*i.e.*: 0.08 kg carbon); however, the wetland can emit 0.04 kg carbon (carbon dioxide and methane) when treating the same amount of wastewater. According to such data, the wastewater of 868,700–956,300 t to be treated yearly can reduce the carbon emission by 34,748–38,252 kg (*i.e.*: about 127–140 t carbon dioxide).

5.4.4.5 About landscape

Rejuvenate and reconstruct multiple types of wetlands, for instance, river wetland (three classes), marsh (forest marsh, grass marsh) wetland, lake wetland, reservoir pond wetland and paddy field wetland to richen the wetland landscape, and make these types of wetland become the propaganda and education base of wetland type and function exhibition, biodiversity (birds, aquatic animals, wetland plants) observation and wetland organic agriculture (organic fishing ponds, organic farmlands).

Bibliography

Gao, W., & Lu, J. J. (2008). A restoration trial of bird habitat on the intertidal flats in the Yangtze Estuary and its short-term effects [J]. *Acta Ecologica Sinica, 28*(5), 2080–2089 (in Chinese).

Guo, W. L., Xia, S. Z., Yuan, X., & Pei, E. L. (2010). Rehabilitation of degraded Dalian Lake wetland in Qingpu District of Shanghai [J]. *Wetland Science and Management, 6*(2), 21–25 (in Chinese).

Hughes, F. M., Colston, A., & Mountford, J. O. (2005). Restoring riparian ecosystems: The challenge of accommodating variability and designing restoration trajectories [J]. *Ecology and Society, 10*(1), 10–12.

Lu, H. W., Zeng, G. M., Jin, X. C., & Jiao, S. (2003). Theories, technologies and applications of ecological restoration and reconstruction on aquatic-terrestrial ecotone [J]. *Urban Environment & Urban Ecology, 16*(6), 91–93 (in Chinese).

Peng, S. L. (2001). The restoration of degraded ecosystems and restoration ecology [J]. *China Basic Science, 3*, 18–24 (in Chinese).

Wu, D., Yue, F., Luo, Z. K., & Wang, T. H. (2011). Ecological effects of lakeside wetlands restoration in Dalian Lake, Shanghai [J]. *Acta Ecologica Sinica, 31*(11), 2999–3008 (in Chinese).

Xu, P. Z., & Qin, B. Q. (2002). Degeneration of ecosystem of lakeside zone around Tai Lake and planning for its rehabilitation [J]. *Water Resources Protection, 3*, 31–36 (in Chinese).

Yan, C. Z., Jin, X. C., Zhao, J. Z., Ye, C., & Wang, Z. Q. (2005). Ecological restoration and reconstruction of degraded lakeside zone ecosystem [J]. *Chinese Journal of Applied Ecology, 16*(2), 360–364 (in Chinese).

Zhang, S., & Tang, Y. J. (1995). *Study on water pollution control for Baiyangdian Lake area (1): Environmental characteristics and management of land/water ecotone [M]* (pp. 42–45). Beijing: Science Press (in Chinese).

Zhang, Y. Z., & Wang, R. (2001). A review of ecological restoration studies on natural wetland [J]. *Acta Ecologica Sinica, 21*(2), 309–314 (in Chinese).
Zhang, H. W., Zhu, X. D., Che, Y., Zhuo, Y. W., & Hu, W. (2009). Approach for ecological restoration of Dalian Lake based on regional development and water source protection [J]. *China Water & Wastewater, 25*(18), 6–9 (in Chinese).
Zhu, X. L., & Yang, Z. Y. (2008). The empirical analysis on the factors influencing the demobilized fishermen in Shanghai Dianshan Lake water source protection areas [J]. *Journal ofAgrotechnical Economics, 3*, 106–112 (in Chinese).
Zhuo, Y. W. (2011). Study on the ecological restoration project of Dalian Lake [J]. *Journal of Anhui Agricultural Sciences, 39*(24), 14799–14800 (in Chinese).

Chapter 6
Wetland Restoration Project Layout and Subarea Design

6.1 General Engineering Layout

The project is the first-stage project of Dalian Lake wetland restoration project (2,000 mu): Shanghai Dianshan Lake Wetland Restoration Demonstrative Project (in the west suburb of Shanghai); the total area of the project zone is about 420,000 m^2 (625 mu), including 60 % fishing ponds and 20 % forest lands, and 20 % lakes, roads and rivers. According to the different conditions of the area and the requirements of wetland restoration, the project area is divided into three functional subareas, *i.e.*: wetland restoration and reconstruction area (Subarea A), forest wetland cultivation area (Subarea B), shoal wetland restoration and biodiversity conservation area (Subarea C). The three subareas involve in different ecological functions and restoration focuses.

6.1.1 Wetland Restoration and Reconstruction Area (Subarea A)

The area of 150 mu in the west of Douji Floodgate is the wetland restoration and reconstruction area (Subarea A), which covers the total area of 101,000 m^2. Now there are 15 fishing ponds of any sizes; the total area of the water surface is 66,000 m^2; the land area is 22,000 m^2; the average bottom elevation of fishing ponds is 1.10 m, with the sediments of 20 cm; the average elevation of pond bank is 3.90 m (referring to the elevation of Wusongkou, Shanghai). There are a small quantity of *Phragmites communis* vegetation in the area, which is about 2,000 m^2; additionally the weeds, dominated by *Gramineae* and *Asteraceae* (*Erigeron annuus*, etc.), grow on pond banks.

The Subarea A shall be rejuvenated the healthy aquatic ecological system and the constructed the ecological organic agriculture. The healthy aquatic ecological

S. An and L. Wang, *Wetland Restoration: Shanghai Dalian Lake Project*,
Springer Environmental Science and Engineering, DOI 10.1007/978-3-642-54230-5_6,

system includes two aspects, namely water purification capacity improvement and aquatic animals and plants system construction.

For the purpose of purifying water, constructing healthy aquatic ecological system and wetland ecological organic agriculture, the Subarea A shall focus on wetland restoration and reconstruction. Firstly, it is to conduct the microrelief rehabilitation on the original fishing ponds to construct them into the continuous and changeable habitats suitable for plants and animals for survival, then launch the plant species introduction engineering and the animal breeding engineering to make them grow into the self-rejuvenating and self-developing healthy ecosystem to improve the sewage purification function and yield function of wetland finally.

For the animal and plant species introduction and breeding, the multiple factors, for instance, the local species, food net structure, economic benefit, biodiversity etc., are considered. After the regional restoration completes, the wetland's pollutant-removing capacity can be improved by the healthy aquatic ecosystem on one hand, and some economic incomes shall be provided to the local residents by the wetland organic agriculture on the other hand.

6.1.2 Forest Wetland Cultivation Area (Subarea B)

Implement the plant replenishment, expansion and replacement engineering on the original wetland forest to make it become the wetland forest cultivation area (Subarea B); such wetland forest cultivation area shall be of 219,000 m^2. There are now 16 fishing ponds and 4 arable lands in Subarea B; where the total area of the fishing ponds is 80,000 m^2; the total area of the forest land is 101,000 m^2; the area of pond banks, roads, rivers and others is 52,000 m^2; the average bottom elevation of pond is 1.40 m, with the sediments of 20 cm; the average elevation of pond bank is 3.30 m. Now the existing plants in Subarea B include 150 mu *taxodium ascendens* forest and the conservation forest formed by *Metasequoia glyptostroboides, Cinnamomum camphora*, as well as a great number of *Phragmites communis* (almost 45 mu) in the coastal shoal.

Forest wetland cultivation area (Subarea B) shall focus on the reconstruction and displantation of the original forest and the watercourse excavation works. The forest expansion is the need of the Subarea B for water conservation; the displantation project is to change the trees unsuitable for growing in wetland habitat; which can also improve the local biodiversity, implement ecological system structure and improve the stabilization of the ecological system.

Due to the original forest and vegetation in the project area are of years and grow well currently, they have already formed a sound forest wetland landscape. In accordance with the principle of "in line with the local conditions", the Subarea B shall focus on the expansion construction of the *taxodium ascendens*. The specific works are to rejuvenate the landform unsuitable for trees for growing in the original wetland forest to make it become the flat bottom suitable for aquatic plants for growing, and then implement the wetland forest expansion project. By replenishing plantation (with the density difference) and replacing the tree species unsuitable for growing in wetland, construct the ecological conservation forest with large area, biodiversity and available for birds for inhabitation and finding foods.

Meanwhile, launch the watercourse excavation project in the Subarea A to connect water systems and enhance water fluidity.

6.1.3 Shoal Wetland Restoration and Biodiversity Conservation Area (Subarea C)

Rejuvenate the area where the water is not deep and becomes the shoal into the shoal wetland and diversity conservation area (Subarea C); which covers the area of 100,000 m^2. In Subarea C there are 15 fishing ponds, where the total area of fishing ponds is 87,000 m^2; the pond bank top area 12,000 m^2 and other area 1,000 m^2; the average bottom elevation of pond is 1.48 m, with the sediments of 20 cm; the average pond bank elevation is 3.90 m; some areas have become the shoal.

For the Subarea C, it is to rejuvenate the landform with the relatively small quantity of engineering and rejuvenate plants by relying on the original habitats suitable for birds. After the Subarea C is rejuvenated, the particular shoal wetland landscape can be formed on one hand, and the wading birds can be attracted to improve biodiversity on the other hand.

In view of the shoal wetland is very beneficial for diversity conservation, the Subarea C shall focus on the restoration of the shoal wetland that is capable of providing habitat for multiple animals. The specific measures are to rejuvenate the microrelief of the original unused fishing ponds and take advantage of the existing natural shoals to construct the continuous and changeable habitat substrate suitable for shoal plants for survival, and then implement the shoal wetland plant restoration project on such habitat substrate. Form different styles of shoals in different waterlines and provide habitats and food source for wading birds, and then provide local habitat diversity and biodiversity.

6.2 Engineering Subarea Design

6.2.1 Design of Wetland Restoration and Reconstruction Area

6.2.1.1 Landform Building and Slop Protection Design

Landform Building

For the wetland restoration and reconstruction area (Subarea A), the Douji Floodgate is in its east; where the bank of the watercourse in the east is the cemented bank, which is long 280 m; the widest reach of the watercourse is 27 m, which is in a place about 38 m away from Douji Floodgate in the south; the narrowest reach is only 7.5 m (between Douji Floodgates); its bank in the south is the silty bank, which is long 365 m; where the widest reach is 19 m (in the southwest of the region) and the narrowest reach is only 8 m (in the place 70 m away from the watercourse in the east). There are 15 fishing ponds of any sizes in the zone; most of them are rectangular irregularly, with the area between 13 m^2 and 9,423 m^2; the banks between ponds are the silty bank.

1. Landform rebuilding of original fishing ponds.

In order to keep the landscape wholeness and the integrity of wetland structure and function, the banks between fishing ponds can be excavated to make all fishing ponds form an integral and up-down water; where the excavation depth is 1.08 m (*i.e.*: the average bottom elevation of fishing ponds). Meanwhile, in order to create multiple deep aquatic habitat systems, some areas can be excavated further by 50 cm, and the other areas can be refilled to lift the height by 1 m.

Additionally, in order to improve the diversity of ecosystem, some submerged hidden banks are constructed (some pond banks are maintained) whilst reforming landform; where the top elevation is 1.60 m, tall 50 cm, long 17 m and wide 12 m. Such reformed landform can meet the different requirements of different aquatic plants on water level and is helpful to form a diversified habitat.

2. Island in lake.

Pile the pond bank soil and the sediment at the bottom of fishing ponds to form two central islands (one large and one small); the island structure and shape can be designed according to the wind direction, water flow and soil distribution conditions.

The design shape of the large central island (Island a) is the proximate arc; where the widest part is 49 m, the narrowest part is only 17 m, and its length is 390 m; in this island, there are four ponds of different sizes. The water inside the island and the water outside the island is connected each other through a small gap, which is wide between 1 and 3 m. Design the slope protection of the inner slope and the external slope of the island as per the slope ratio of 1:3 and 1:4 respectively, and then design the different vegetation zones accordingly. Provided it reaches the maximal water level, *i.e.*: 2.68 m, the area that "Island a" comes out from the water surface is 6,367 m^2; however, when it reaches the minimal water level, *i.e.*: 2.08 m, such area can be 9,932 m^2; but when the water level is 1.08 m, such area is 15,764 m^2.

In accordance with the ecological purification requirements of the area, the small central lake (Island b) is designed into a quick-infiltration ecological system, which is formed by excavating the sediments. The quick-infiltration ecological system can efficiently relieve the pollution of area, provide beneficial conditions to healthy aquatic ecosystem, "digest" the sediment removed from the fishing ponds in the area and decompose the pollutants in some sediments. The specific measures are: excavate about 8,300 m^3 sludge from the fishing ponds in the area, and then compress and solidify the sediments; the sediments can be solidified and compressed into a volume of 2,800 m^3 around; afterwards, refill these compressed sediments into the quick-infiltration ecological system. The specific structure of such quick-infiltration ecological system is: pile 25 cm gravel layer (particle diameter between 2 and 4 mm) and 20 cm coal cinder layer (diameter 0.8–1.2 mm) on the silty layer in turns, and then cover the pervious cloth of 1,100 m^2 on the coal cinder layer; afterwards refill the soil and new soil on the pervious cloth by 205 cm and 30 cm respectively; such structure can efficiently relieve the release speed of pollutants in sediments. The quick-infiltration ecological system adopts the ecological bags around its slopes, which are placed at its bottom, such way can also rejuvenate some submerged plants; on the slope, the ecological protective net is placed, which can help plants for growth. Under the high water level of 2.68 m currently, the area reaches 1,270 m^2; at the water level of 1.08 m, the area is 1,750 m^2. The slope ratio for the elevation less than 2.68 m is 1 : 3, which guarantees the stable slope and plant structure.

Whilst reforming the landform of the pond banks, pond bottom and central islands, keep earthwork excavation and refill balance; no spoil shall be produced.

Slope Protection Design

In accordance with the landform conditions and habitat diversity of the area, as well as the principle to keep slope stable and safe, the slope in this area shall be designed three different slope ratio, *i.e.*: 1:7, 1:4 and 1:3. Among which, the slope with the first ratio is centralized in the east and north of the area; where the east slope is 280 m long, the north slope is 350 m long; the east slope's area is 3,500 m^2, the north slope's area is 2,550 m^2; the slope with the second ratio is distributed in the place outer of the large island, which is long 900 m and covers an area of 3,500 m^2; the slope with the third ratio is mainly distributed in two sections, some are in the internal side of the large islands, which is long 750 m; some are in the periphery of the small island, which is long 150 m; the former covers an area of 1,530 m^2 and the latter one covers an area of 750 m^2.

6.2.1.2 Plant Configuration

For the project, the plants are selected by following the principles as follows, namely the plants are benefit for the wetland to exert its purification function, benefit for biodiversity, as well as the principle in line with the local conditions. Meanwhile, the landscape effect and the seasonal aspect change should be considered.

According to the difference of slope ratio and along the direction from the island towards the water surface, design the area into the forest wetland zone (including: *Pterocarya stenoptera*, *Metasequoia glyptostroboides*, *Sapium sebiferum*, etc.), bushwood wetland zone (including *Rosa multiflora*, *Forsythia viridissima*, *Cyperus rotundus*, etc.), emergent aquatic plant zone (including: *Phragmites communis* and *wild Zizania caduciflora*) and the submerged plant zone (including *Hydrilla verticillata*, etc.); add the floating leaved plants (including *Euryale ferox Salisb.*, etc.) in the bays of islands. For the specific layout, the configuration sequence of the two islands, from up to down, is the forest wetland zone, bushwood wetland zone, emergent aquatic plant zone and submerged plant zone. Plant the floated-leaved plants in the bays of islands; plant emergent aquatic plants and submerged plants on the surrounding slopes of the project, and plant the submerged plants in the pond banks and the upward refilled areas.

The trees are *Ulmus pumila, Sapium sebiferum, Salix matsudana, Pterocarya stenoptera, Albizia julibrissin, Metasequoia glyptostroboides, Ascendens mucronatum, Taxodiumascendens, Glyptostrobus pensilis, Taxodium distichum*, etc.; on the tree ground, the *Commelina communis, Hemerocallis citrina, Iris japonica, Ajuga multiflora, Acorus gramineus* and other sciophilous and hygrophilous herbs can be planted. Since the special soil conditions, the coastal low wetland zone should be planted the deciduous tree (the principal tree) and other wild flowering ground cover plants so as to form the natural coastal vegetation zone and highlight the landscape effect of the seasonal aspect color and trees reflections in water. Additionally, according to the local water depth, the different plants can be planted so as to form the multi structure and multi-hierarchical vegetation zones.

In accordance with the site conditions of plants, the slope ratio of the local structure and the design requirements of the local area, the ecological revetment is designed as follows:

1. The pond bank preservation zone shall only be planted the submerged plants; this is the submerged vegetation zone and it covers the area of 300 m^2; where the plantation density is 10 plants/m^2 and the 1–1.5 m submerged plants shall be planted; after the local water environment turns better and the water transmittance enhances, the existing submerged plants can be spread as the species source to form a good submerged plant format.
2. The surroundings of the project area shall only be planted the emergent aquatic plants and the submerged plants, which coves an area of 6,000 m^2. Among which the former covers an area of 5,000 m^2; the latter covers an area of 1,000 m^2. The emergent aquatic plants are the *Phragmites communis* and the aquatic vegetables with the economic benefit, which can purify water quality, but also help to get economic benefits.
3. According to the design for the large island (Island a) and the small island (Island b), the internal water areas shall have the forest wetland zone, bushwood wetland zone, emergent aquatic plant zone and submerged plant zone, they cover an area of 300 m^2, 1,000 m^2, 450 m^2 and 247 m^2 respectively. Corresponding to the sediment used as the backfill material, the forest plants and the bushwood plants in the area can absorb the heavy metal in sediments; whereas the emergent aquatic plants and the submerged plants can absorb total nitrogen, total phosphor and other pollutants; all of them play the role of the quick-infiltration system in water purification.
4. For the outer area of the large island (Island a) (without bays), the forest wetland zone, bushwood wetland zone, emergent aquatic plant zone and submerged plant zone are designed, which covers an area of 2,950 m^2, 2,410 m^2, 10,000 m^2 and 1,800 m^2 respectively. The plantation of multiple plants makes huge contribution to the improvement of biodiversity and the absorption of pollutants.

5. The outer area of the large island (Island A) (with bays) are distributed the forest wetland zone, bushwood wetland zone, emergent aquatic plant zone, floating-leaved plant zone and submerged plant zone. Comparing with that of the outer area without bays, the water in this area is stable relatively, the wind speed is slow relatively, therefore the floating-leaved plants can be planted. The floating-leaved plants can not only enhance purification capacity, but also improve the biodiversity and produce certain economic and landscape value. The plantation area of the floating-leaved plant is 750 m^2.

All plant zones are not separated completely, but are intercrossed each other; all of them form an integrated vegetation zone.

6.2.1.3 Establishment Healthy Aquatic Ecosystem

The key point of the water quality recovery is to establish the healthy aquatic ecosystem; the water quality of the natural water area is of self-regulation, the rainfall replenishes the water quantity, and the chemical nature of water (for instance, pH value and hardness) is subject to the demonstration and the minerals of the ground surface where river passes through; the flow of water makes water's surface change from time to time. The existing Dalian Lake fishing pond is a relatively closed ecosystem with less volume. Due to the lack of the aquatic plants and the bank vegetation, as well as the simple slope construction format of artificial fishing pond and original breeding specifies, the local food chain and food net can not be established efficiently; therefore the existing pollutants can not achieve the self-cleaning and degradation in the local fishing ponds, but accumulate slowly and result in the breakdown of the fishing pond ecosystem finally.

The first job to construct the healthy aquatic ecosystem is to adjust the plant type and the plant structure in water. In this design, on one hand, plants are planted on banks, including the trees, bushes, and the grass; where the trees and bushwood can absorb the heavy metal in soil; the grass can absorb other nutritive elements; on the other hand, the emergent aquatic plants are planted on banks, including *Phragmites communis*, Spiked Loosestrife, *Zizania caduciflora* and *Acorus calamus*, which can absorb the nutritive substances in water to reduce the concent of TOC, TN and TP obviously; meanwhile, the submerged and floating plants planted in water can also make contributions to the restoration of the healthy aquatic ecosystem.

After completing plant configuration and restoration, the job to introduce and rejuvenate animals can be launched. According to the "energy tower" principle and the substance flow principle of food chain and food net, the project is designed the corrosive, herbivorous and predacious fishes and other aquatic animals; additionally, some birds can be attracted; for the efficient healthy aquatic ecosystem, see the relationship as Fig. 6.1.

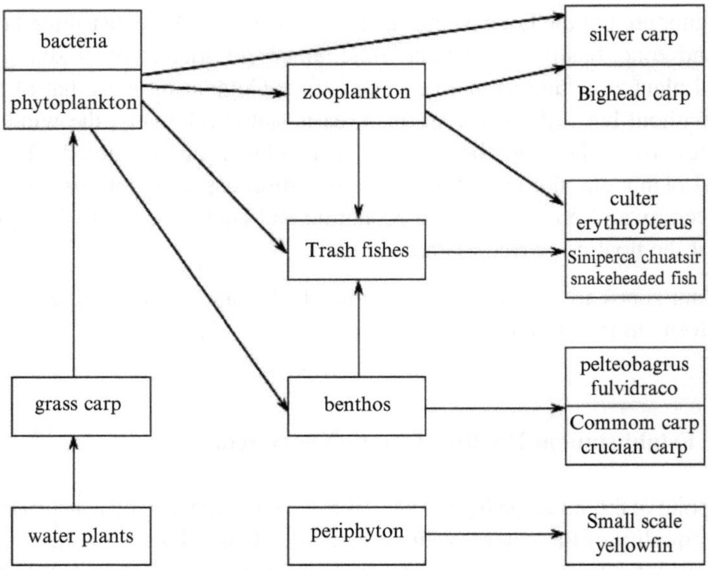

Fig. 6.1 Establishment of healthy food chain and food net

In accordance with the Dalian Lake water and food chain structure, the grass carp and the Wuchang fish eat water plants; the silver carp eats the phytoplankton; the bighead carp eats zooplankton; the *Tilapia, sparus latus, Pelteobagrus fulvidraco*, common carp and crucian carp eat larva of insects, aquatic higher plant debris, trashes and algae and other omnivorous food; the elopichthys bambusa Richardson is predacious fish and natural disposition is fierce. In accordance with the food characteristics of the fishes above, these fishes can be utilized each other in the breeding process; for instance, the grass carp eats the water plants, its excrement can fertilize the water and help plankton to breed and grow, and then which can be ate by silver carp and the bighead carp; however, after the silver carp and the bighead carp eats the plankton, the fertility of water is reduced, which is helpful for grass carp to grow; the predacious fishes can reduce the breeding of trash fishes; therefore the additional value and the yield rate of the aquatic products can then be improved by breeding the different species of wild fishes and establishing rational aquatic ecosystem; which achieves not only the anticipation on wetland restoration and water pollution governance, but also compensates the economic loss arising from "return the ponds to wetland" (Fig. 6.1).

The landform reconstruction and the vegetation restoration project of the project and the area can not be separated completely; it is needed to add and breed aquatic animals, for instance, fishes, shrimps, crabs, etc. on the reconstructed habitat base. For the plant species and animal quantity, see Table 6.1.

Table 6.1 Quantity of plants and animals in Subarea A

Species	Unit	Quantity	Size
Pterocarya stenoptera	Plant	50	Diameter at breast height: 10 cm
Taxodium ascendens	Plant	100	Diameter at breast height: 10 cm
Taxodium distichum	Plant	50	Diameter at breast height: 12 cm
Papermulberry	Plant	50	Diameter at breast height: 12 cm
Metasequoia glyptostroboides	Plant	400	Diameter at breast height: 12 cm
Salix matsudana	Plant	50	Diameter at breast height: 12 cm
Sapium sebiferum	Plant	50	Diameter at breast height: 12 cm
Glyptostrobus pensilis	Plant	50	Diameter at breast height: 12 cm
Morus alba	Plant	50	Diameter at breast height: 12 cm
Melia azedarach	Plant	50	Diameter at breast height: 12 cm
Alnus trabeculosa	Plant	50	Diameter at breast height: 12 cm
Albizia julibrissin	Plant	50	Diameter at breast height: 12 cm
Osmanthus fragrans	Plant	50	Diameter at breast height: 12 cm
Bambusa glaucescens	Plant	100	Tall: 100 cm
Salix purpurea	Plant	100	Tall: 100 cm
winter jasmine	Plant	100	Tall: 100 cm
Jasminum floridum	Plant	100	Tall: 100 cm
Lonicera japonica	Plant	100	Tall: 100 cm
Forsythia viridissima	Plant	100	Tall: 100 cm
Buddleja lindleyana	Plant	100	Tall: 100 cm
Distylium buxifolium	Plant	100	Tall: 100 cm
Weeping forsythia	Plant	100	Tall: 100 cm
Rosa multiflora	Plant	100	Tall: 100 cm
Polygonum orientale	Plant	400	Tall: 100 cm
Physalis alkekengi	Plant	100	Tall: 100 cm
Pennisetum alopecuroides	Sq.m	500	Grass seed sows. grass and sod half-half
Setaira viridis	Sq.m	1,000	Grass seed sows. grass and sod half-half
Carex tristachya	Sq.m	500	Grass seed sows. grass and sod half-half
Schoenoplectus trigueter	Sq.m	500	Grass seed sows. grass and sod half-half
Cynodondactylon	Sq.m	500	Grass seed sows. grass and sod half-half
Mat grass	Sq.m	500	Grass seed sows. grass and sod half-half
Nutgrass flatsedge	Sq.m	500	Grass seed sows. grass and sod half-half
Typha orientalis	Plant	500	Young plant, plant height: 20 cm
Acorus calamus	Plant	500	Young plant, plant height: 20 cm
Softstem bulrush	Plant	500	Young plant, plant height: 20 cm
Ludwigia peploides Subsp. Stipulacea	Plant	500	Young plant, plant height: 20 cm
Phragmites communis	Plant	60,000	Young plant, plant height: 40 cm
Arundo donax	Plant	2,000	Young plant, plant height: 20 cm
Cortaderia selloana	Plant	2,000	Young plant, plant height: 20 cm
Wide Zizania caduciflora	Plant	5,000	Young plant, plant height: 20 cm
Oenanthe javanica	Plant	5,000	Young plant, plant height: 10 cm
Sagittaria sagittifolia	Plant	3,000	Young plant, plant height: 10 cm
Eleocharis dulcis	Plant	3,000	Young plant, plant height: 10 cm
Alisma plantago-aquatica	Plant	2,000	Young plant, plant height: 20 cm

(continued)

Table 6.1 (continued)

Species	Unit	Quantity	Size
Ipomoea aquatica Forsk	Plant	4,000	Young plant, plant height: 10 cm
Lotus	Plant	1,500	Young plant
Eomecon chionantha	Plant	1,500	Young plant
Hydrocharis dubia	Plant	1,000	Young plant
Pumilum	Plant	1,000	Young plant
Furctus Trapae Quadricaudatae	Plant	1,000	Young plant
Euryale ferox Salisb	Plant	1,000	Young plant
Potamogeton distinctus	Bushwood	10,000	Young plant
Potamogeton crispus	Bushwood	20,000	Young plant
Vallisneria natans	Bushwood	20,000	Young plant
Myriophyllum spicatum	Bushwood	20,000	Young plant
Hydrilla verticillata	Bushwood	20,000	Young plant
Silver carp	kg	1,600	30 fishes/kg
Bighead carp	kg	1,600	30 fishes/kg
Crucian carp	kg	400	30 fishes/kg
Grass carp	kg	1,250	50 fishes/kg
Channa argus	kg	150	30 fishes/kg
Common carp	kg	900	30 fishes/kg
Siniperca chuatsi	kg	150	30 fishes/kg
Parabramis pekinensis	kg	1,000	30 fishes/kg
Elopichthys bambusa	kg	150	30 fishes/kg
Catfish	kg	900	30 fishes/kg
Pelteobagrus fulvidraco	kg	400	30 fishes/kg
Channa maculata	kg	500	30 fishes/kg
Shrimp seed	kg	500	200/kg
Crab seed	kg	100	20/kg

6.2.2 *Design of Forest Wetland Cultivation Area*

6.2.2.1 Landform Building and Slope Protection Design

Landform Building

For the forest wetland cultivation area (Subarea B), the Weinei Xiaohebang is in the east, Dalian Lake is in the west, Lanlu Port is in the north and Subarea C is in the south. Currently there are 16 fishing ponds of different sizes in the area, as well as the *Metasequoia glyptostroboides* forest, *Taxodiumascendens. Brongn* forest and *Cinnamomum camphora* (L.) forest of almost 100,000 m^2; the landform of the area looks like scattered and the ecological landscape is not continuous. Among which there are *Metasequoia glyptostroboides* forest of 32,000 m^2, *Taxodiumascendens. Brongn* forest 50,000 m^2 and *Cinnamomum camphora* forest 18,000 m^2; where the fishing pond covers an area of 615–17,736 m^2; the sediment depth is around 20 cm.

In order to ensure the integrity of landscape and wetland structure and function, the engineering design proposes to fill all fishing ponds in the area to form a relatively flat terrestrial landform. After rejuvenating, the original uneven landform with multiple pond banks becomes a flat landform suitable for plantation; its average elevation is very similar to that of the existing *Metasequoia glyptostroboides* forest and is lifted slightly (2.7 m) so as to ensure the security of plantation. On the land, the different vegetation zones are designed as per the preferred density ratio.

Additionally, in view of the relatively simple geomorphology of the forest wetland, silted rivers and blocked water system, it is proposed to excavate the local area and construct fishing ponds, as well as dredge rivers to construct the 4-cascading rivers. For the specific renewal measures, see "Watercourse Adjustment".

Slope Protection Design

The new waterways in the area shall be designed the ecological slope, which is a slope with the slope area of 1:2.5; where the watercourse is a 4-cascading watercourse (R1, R2, R3, R4); with the width 11 m, 9 m, 7 m and 5 m respectively; varied striking plants shall be planted at the slopes at both sides of the watercourse to distinguish watercourse grade and create different vegetation landscape.

6.2.2.2 Plant Configuration

For the project, the plants are selected by following the principles as follows, namely the plants are benefit for the wetland to exert its purification function, benefit for biodiversity, as well as the principle in line with the local conditions. Meanwhile, the landscape effect and the seasonal aspect change should be considered.

The existing *Metasequoia glyptostroboides* forest and *Taxodiumascendens. Brongn* forest in the project area are preserved well; however, the original *Cinnamomum camphora* (L.) Presl forest, since they do not bear water logging and the local underground water level is relatively high, most of them have already dead. After the project is implemented, the water conditions shall be changed; so it is estimated the existing *Cinnamomum camphora* (L.) Presl forest shall have no way to survive; therefore the *Ascendens mucronatum*, since it is bearable the water and has landscape and ecological effect, shall be selected to replenish.

As per the original tree forest, the area is divided into three subareas, *i.e.*: Subarea A, Subarea B and Subarea C. The "Subarea A" is an area centered by the original tree forest; the "Subarea B" is the area outside the Subarea A; the "Subarea C" is the area outside the Subarea B. The original *Cinnamomum camphora* forest, after being replaced by the *Ascendens mucronatum*, shall form the *Ascendens mucronatum* forest. In Subarea A, the *Taxodiumascendens*, *Metasequoia glyptostroboides* and *Ascendens mucronatum* shall be planted with the plantation gap of 2.5–4.0 m; the plantation area of such three tree species shall be 42,000 m^2, which shall cover 5,500 trees totally. Plant these tree species randomly at the given plantation gap to form a relatively

Table 6.2 Quantity and size of plants in Subarea B

Species	Unit	Quantity	Size
Metasequoia glyptostroboides	Plant	8,500	Diameter at breast height: 10 cm
Taxodiumascendens	Plant	5,000	Diameter at breast height: 10 cm
Ascendens mucronatum	Plant	6,000	Diameter at breast height: 10 cm
Nymphaea alba	Plant	5,000	Young plant
Euryale ferox Salisb	Plant	5,000	Young plant

natural status. In Subarea B, the *Taxodiumascendens*, *Metasequoia glyptostroboides* shall be planted with the plantation gap of 3.5–5.0 m; the plantation area of such two tree species shall be 27,900 m^2, which shall cover 1,760 trees totally; the plantation mode in this area is similar to that of the Subarea A; the Subarea C shall be planted the *Taxodiumascendens*, *Metasequoia glyptostroboides* with the plantation gap of 4.5–7.0 m; the plantation area of such two tree species shall be 34,500 m^2, which shall cover 1,360 trees totally; the plantation mode in this area is similar to that of Subarea A and B. In such way, a landscape with the plantation density reducing from centre to surrounding shall be formed, which can not only offer more free growth space for plants, but also connect with the wetland restoration area perfectly.

Meanwhile, the aforesaid forests shall be excavated the waterways to divert water into the wetland forest to form particular wetland forest landscape, but also make the water system in the whole area open each other to provide convenience for the future safekeeping and supervision works. The watercourse shall be divided into four cascades, namely 11, 9, 7 and 5 m wide; the watercourse of every cascade shall be planted different plants at both sides, namely the trees, bushwood, emergent aquatic plants and floating-leaved plants. The different cascades of watercourses shall be available to see the different plant landscapes.

Additionally, in order to coordinate the diversity of ecological environment and landscape, ten areas of different sizes in the forest wetland are selected to plant the aquatic plants. Some area are planted by taking advantage of the original fishing ponds; some areas are planted directly on lands; some areas are planted by excavating the shallow water ponds; which shall achieve the maximal landscape and ecological effect with the minimal engineering quantity.

For the plant species and animal quantity, see Table 6.2.

6.2.3 Design of Shoal Wetland Restoration and Diversity Conservation Area

6.2.3.1 Landform Building

For the Subarea C, the Weinei Xiaohebang is in its east, Dalian Lake is in the west, and the Subarea B is in its north; its south is connected with the second-stage wetland project. The north bank of the area is 532 m long, the south bank is 527 m long, and the west bank is 340 m long; there are five fishing ponds of different size (between 519 and 12,505 m^2) in the area, with the sediment around 20 cm. Since the land was transferred, some fishing ponds form the shoal wetland landform gradually; additionally there are plenty of birds for habitation. However, since the pond banks separate the ponds each other and the original building intervenes, the habitat in this area is not continuous.

In order to guarantee the integrity of the landscape wholeness, wetland structure and function, the design is proposed to level and flatten the banks between fishing ponds and make submerged landform reconstruction so as to form a relatively flat and partially up-down submerged landform to facilitate the shoal wetland to rejuvenate. Meanwhile, in order to build the diversified submerged landform, some banks are preserved in the area; some banks shall be lowered to plant various terrestrial plants and aquatic plants. The total length of the bank preserved is 320 m, with the width around 5–6 m. For the bank beneath the high water level, the slope ratio of 1:25 is adopted to make the submerged reconstruction. Due to the slope ratio of the original fishing pond is between 1:2 and 1:3, the submerged part of the area does not need to change too much; however, on the land, the vegetation zones shall be designed as per the given density ratio.

For the landform map before and after reconstruction, see the attached picture. From which it can be seen the landform becomes flat and goes up and down regularly after being reconstructed; which lays sound base for the shoal wetland. In order to guarantee the landscape of the shoal wetland, the wetland of the regular height is specially designed to ensure an area of 20,000 m^2 to come out of the water surface when it is at the high water level of 2.68 m, *i.e.*: over 20 % shoal landform emerges and an area of 60,000 m^2 to come out of the water surface when it is at the low water level of 2.08 m, *i.e.*: almost 60 % shoal wetland emerges, and then show the characteristics of the shoal wetland; such reconstruction shall produce profound impact on the ecological environment of the shoal. In order to achieve the effect, it is planned to excavate the bank by 80,000 m^2 and excavate by 3.9 m deeply.

6.2.3.2 Plant Configuration

After the landform reconstruction completes, the plant plantation work can be launched. In this area, the plant restoration shall focus on the natural restoration mainly, but also include the aquatic emergent plants and the floating plants with tuber so as to provide foods for wading birds and fishes. In the portion where comes out from the water surface when it is at the high water level, the *Taxodiumascendens. Brongn*, *Metasequoia glyptostroboides* and *Ascendens mucronatum* shall be planted, which can not only protect the submerged parts of the shoal, but also correspond to the forest wetland zone and the wetland restoration zone to enhance regional landscape effect. In the restoration project of the area, it shall focus on establishing the healthy aquatic ecosystem based on the diversified aquatic plants and aquatic animal population.

Table 6.3 Quantity and size of plants in Subarea C

Species	Unit	Quantity	Size
Taxodium ascendens	Plant	50	Tree, diameter at breast height: 12 cm
Metasequoia glyptostroboides	Plant	50	Tree, diameter at breast height: 12 cm
Ascendens mucronatum	Plant	50	Tree, diameter at breast height: 12 cm
Osmanthus fragrans	Plant	50	Tree, diameter at breast height: 12 cm
Forsythia viridissima	Plant	50	Bushwood: 100 tall
winter jasmine	Plant	50	Bushwood: 100 tall
Lonicera Japonica	Plant	50	Bushwood: 100 tall
Polygonum orientale	Plant	50	Bushwood: 100 tall
Phragmites communis	Plant	200	Young plant: 40 cm
Furctus Trapae Quadricaudatae	Plant	50	Young plant
Euryale ferox Salisb	Plant	50	Young plant
Hydrocharis dubia	Plant	50	Young plant
Pumilum	Plant	50	Young plant

For the plant species and animal quantity in Subarea C, see Table 6.3.

6.3 Water Conservation and Earthwork Design of the Project

6.3.1 The Work Conservation Design of the Project

6.3.1.1 Project Class and Design Standard

Project Class and Building Class

The scope of the project belongs to the water system within dykes. The project is class III project in accordance with the provisions of the article 2.1.1 of *Classification of Water Conservation and Aquatic Electricity Projects and the Standard of flood* (SL252—2000). The dykes outside the wetland areas belong to class III water conservation buildings; the newly-built rivers belong to class IV water conservation buildings.

Standard for Flood Drainage

In accordance with Water Conservation Planning, the standard for flood drainage of the local area adopts the standard for flood drainage of 180.2 mm rainfall of the longest time 24 h that was encountered once every 20 years like the rainfall in 1963. Of them, the heaviest storm rainfall in an hour is 44.3 mm.

Table 6.4 Design water level of project

Item	Water level/m	Remarks
High design water level	3.50	High flood drainage water level
Normal water level	2.50	Normal water level
Low design water level	2.00	Pre-fall water level

Anti-seismic Standard

Pursuant to geological investigation reports, the seismic fortification intensity of the place where the project is 7°; the designed basic acceleration of ground motion is 0.10 g; subornative design earthquake grouping is group I; foundation soil belongs to soft soil; site classification is class IV.

Design Water Level

The internal rivers of the project are connected with the rivers of Douji Port. So the design water level of wetland adopts the water level values of Douji Port planning. Details are seen in Table 6.4.

Dyke Top Elevation

1. Dykes.
   The dykes outside the wetland of the project belong to class III water conservation buildings. Their safe super height can be 0.7 m (not allow to exceed waves) or 0.4 m (allow to exceed waves) in accordance with *Design Specifications of Dyke Projects*. In consideration of the fact that after the project exceeds waves, it will not cause serious damage to its back ground, the safe super height can be allowed to exceed waves. The high design water level is 3.5 m, and the top elevation of the dykes outside wetland should not be lower than 3.90 m. In overall consideration of the fact that the planning dyke top elevation of the rivers of Douji Port, etc., is 4.00 m, the dyke top elevation outside the wetland of the project is determined to be 4.00 m. The areas whose ground is above 4.00 m can maintain the status quo.
2. The newly-built internal rivers of the wetland.
   The ground elevation of the place where the newly-built rivers of the wetland of the project is about 3.00 m. Because the plants planted within the wetland all can be inundated, namely the water surface of the newly-built rivers can submerge its dykes at high water level, the safe super height and other factors are not considered for the newly-built rivers of the project. The dyke top elevation of the internal rivers of the wetland can change in accordance the need of the overall layout of the wetland and the elevation fluctuation of the terrain that has been transformed.

6.3.1.2 Project Design Principle

The following principles shall be taken into account in the overall layout of the project:

1. The requirements of related overall planning and professional planning, such as the overall planning of Dalian Lake wetland, the water conservation planning of Jinze Town, etc.
2. Under the premise that the dykes outside the wetland meet the requirements of flood protection and drainage, efforts should be paid to multifunction design, such as ecological slope protection and green space, etc., to integrate natural ecological concepts into water system project design.
3. Emphasis should be given to both storage and discharge in the project of the newly-built rivers of the wetland; the plane extension of the rivers can be wider or narrower to embody their natural feature.
4. The side slopes of the rivers mainly adopt the flexible structure of plant slope protection to highlight their natural ecological characteristics.

6.3.1.3 The Water Conservation Project Design of the Wetland

The Plane Layout of the Water Conservation Project of the Wetland

1. The plane layout of the newly-built internal rivers of the wetland.

In light of the relatively single terrain of the forest wetland of B, C areas in which there are sludge-blocked rivers and blind roads, class I, II, III, IV rivers will be constructed within B, C areas in this design, relatively called as R1, R2, R3, R4, so that the internal water systems within B, C areas can be interconnected. R1 as main river will run through B, C areas starting from the confluence of the east bank of B area and existing rivers and streams, ending with the confluence of the east bank of C area and existing rivers streams; its stake No. is R0+000.00~R1+589.85 m, and length 1,589.85 m. R2, R3, R4 are mainly the branches R1; their estuaries are respectively 9 m, 7 m, 5 m; and their total length respectively 1,586.7 m,659.2 m,238.0 m.

2. The plane layout of the dykes.

At present, there are mainly the ridges of fishponds and scattered dilapidated houses of residents around the project area. The dykes around A area are east dyke, west dyke and north dyke; the stake No. is between A0+000.00~A0+834.13 m. The south dyke is connected with the second stage project. So it is ok to retain the elevation of the existing fishponds. The project areas of B, C area are integrated as a whole, and their dykes can be uniformly taken into account. Their dykes are mainly east dyke, west dyke and north dyke; their stake numbers are respectively B0+000.00~B1+439.09 m and C0+000.00~C0+424.33 m; the south dyke is connected with the second stage project. So it is ok to retain the elevation of the existing fishponds. It is necessary to reserve estuaries for the dyke to guarantee the connection of the water systems within the wetland and the existing rivers and streams. It is projected that the length of the dykes around A area is 834 m; the length of the dykes around B area is 1,439 m; the length of the dykes around C area is 943 m.

The Section Design of the Water System Project

1. The section design of the newly-built rivers of the wetland.

The project is mainly ecological wetland restoration; the river section is designed mainly in accordance with the requirements of the wetland restoration to fully take advantage of the functions of the overall wetland environment after river improvement. In the section design, the rivers within the wetland run relatively slowly. Hence, the section of dyke protection adopts earthen side slope with gradient 1:2.5. The dykes top elevation can change depend on transformed terrain elevation fluctuation.

R1 is the main river, its bottom elevation is designed to be 1.00 m below low water level, its bottom width is 3 m, its side slope gradient is 1:2.5, its dyke top elevation is between 3.30 and 2.00 m, its estuary width is 14.5–8 m, and its length is 1,589.9 m.

R2, R3, R4 are mainly the branches of R1. R2 bottom elevation is designed to be 0.40 m, 0.60 m below low water level, its bottom width is 2 m, its side slope gradient is 1:2.5, its dyke top elevation is between 3.30 and 2.00 m, its estuary width is 11.5–5 m, and its length is 1,586.7 m. R3 bottom elevation is designed to be 2.0 m the same with low water level, its bottom width is 2 m, its side slope gradient is 1:2.5, its dyke top elevation is between 3.30 and 2.00 m, its estuary width is 4.5–2 m, and its length is 659.2 m. R4 bottom elevation is designed to be 2.20 m, 0.3 m below normal water level, its bottom width is 2 m, its side slope gradient is 1:2.5, its dyke top elevation is between 3.30 and 2.20 m, its estuary width is 4.75–2 m, and its length is 238.0 m.

2. The section design of the dykes.

The dykes outside the wetland adopt earthen gentle slope so as to be helpful for the growth of aquatic plants and the restoration of ecological environment. They can be the form of gentle slopes with different heights in accordance the characteristics of each part. Pursuant to the requirements of wetland terrain transformation, the dyke gradient can be divided into two kinds, 1:4 and 1:7. The gradient of the east dyke and north dyke of A area is 1:7, their length is all 834.13 m. The gradient of the other dykes of A area is 1:4, their length is all 1,863.42 m. The bottoms of the dykes are naturally connected with the transformed terrain of the wetland areas. The dyke top elevation is determined to 4.00 m in accordance with the requirements of local flood protection and drainage and in association with the dyke top elevation surrounding internal rivers. If lower than 4.00 m, it is necessary to fill it to 4.00 m with soil; if more than 4.00 m, the dyke slope and the original ground shall be naturally connected.

Table 6.5 Safety coefficient table of overall stability and *design norms for embankment project*

Loading combination	Safety coefficient
Normal operating condition	1.20
Abnormal operating condition	1.10

Overall Stability Checking Calculation of River Slopes

The design is for the terrain transformation in Zone C, the partial zones will excavate up to −3.90 m elevation, and the slope stability of the most unfavorable section is calculated in this time by referring to the geological information of Dalian Lake and communication projects of surrounding water systems. Because the slope of the outer embankment is gentle and slope stability is good, they will not be calculated in this design. The cross section of the inland river embankment of the wetland is same as Zone C, and it will not be calculated in this time.

Calculation of working conditions is divided into three types, namely perfect completion conditions, basic combination and special (earthquake) combination, the groundwater level is below 1 m, the channel water level is water-free (perfect construction works), 2.00 m (design low water level), and 2.50 m (normal water level). The fair values of safe coefficient are shown in Table 6.5.

1. According to Article 9.2.4 of Basic Design Norms for Foundation (DGJ08—11—1999), the stability checking calculation of the most dangerous circular sliding surface can adopt the following formula in the simple slice method (Swedish method) when the shear strength of the earth body is confirmed in the total stress method:

$$\gamma_0\gamma_s\sum_i(q_{ki}b_i+w_{ki})\sin\alpha_i\le\frac{1}{\gamma_R}\left[\sum_i c_{ki}L_i+\sum_i(q_{ki}b_i+w_{ki})\cos\alpha_i\tan\varphi_{ki}\right]$$

When seismic checking is needed, it can be calculated according to the following formula:

$$\gamma_0\gamma_s\sum_i(q_{ki}b_i+w_{ki})\sin\alpha_i+Q_{ki}\frac{y_i}{R}\le\frac{1}{\gamma_R}\left[\sum_i c_{ki}L_i+\sum_i(q_{ki}b_i+w_{ki})\cos\alpha_i\tan\varphi_{ki}\right]$$

In which q_{ki} is the normal loading value acting on the *i*th top surface of earth stripe (kN/m^2); b_i is the width of the *i*th earth stripe; w_{ki} is the deadweight standard value of the *i*th earth stripe (kN/m); the earth body below the underground water level shall adopt buoyant gravity density for calculation when there is no seepage action; α_i is the angle between the central tangent and the horizontal line of the sliding surface arc of the *i*th earth stripe (°); c_{ki} is the standard value of the soil cohesion on the sliding surface of the *i*th earth stripe (kPa); φ_{ki} is the standard value of the internal friction angle of the soil on the sliding surface of the *i*th earth stripe (kPa); L_i is the arc length of the cohesion the sliding surface of the *i*th earth stripe (m) and $L_i = b_i / \cos\alpha_i$; y_i is the vertical distance from the center of gravity to the circular center of the sliding arc of the *i*th earth stripe (m); R is the radius of the sliding arc (m); Q_{ki} is the normal value of the horizontal seismic force of the *i*th earth stripe (m); when the seismic fortification intensity is 7°, $Q_{ki} = 0.025(q_{ki} b_i + w_{ki})$; γ_s is the partial coefficient for action and it is 1.0; γ_0 is the structure importance coefficient for, it is 1.0 for the general slope project; and it is 1.1 for the slope project; γ_R is the partial coefficient for resistance, it is 1.3 for the design calculation, it is 1.25 for the checking calculation, and it is 1.0 for the seismic calculation.

2. According to *Design Norms for Embankment Project*, the stable analysis adopts Swedish slice method. The calculation formula of the ant-sliding safety coefficient is shown as follows.

   The construction period:

$$K = \frac{\Sigma\left(C_u b \sec\beta + W\cos\beta\tan\varphi_u\right)}{\Sigma W\sin\beta}$$

The operation period:

$$K = \frac{\Sigma\left[C_{cu} b\sec\beta + \left(W\cos\beta - u_i b\sec\beta\right)\tan\varphi_{cu}\right]}{\Sigma W\sin\beta}$$

$$W = W_1 + W_2 + \gamma_w Zb$$

Formula: b is the width of the slice (m); W is the gravity of the slice, $W = W_1 + W_2 + \rho_w Zb$ (kN); W_1 is the gravity of the slice above the water level of the outer embankment (kN); W_2 is the gravity of the slice below the water level of the outer embankment (kN); Z is the distance of the water level of the outer embankment which is higher that the bottom surface of the slice (m); u is the pore pressure of the embankment body or foundation in the stable percolation period (kPa); u_i is the pore pressure of the embankment body or foundation before the fall of the water level (kPa); β is the angle between the gravity line and the radius crossing the center of its bottom surface (°); γ_w is the weight of the water (kN/m^3); C_u, φ_u, C_{cu}, φ_{cu} are the indexes of the shear strength of the earth (kN/m^3, °).

Through the calculation achievements, the overall stability of the channel slopes of the project satisfies with the standard requirements, the gradient of the channel slop is less than 1:25, the completion period and operation period are safe. In the next phase design, they shall be checked according to the geological conditions of the different parts and the geological survey information of the project (Table 6.6).

Table 6.6 Achievement table of overall stability calculation of river channel slopes

Project item	Calculation of working conditions	Safety coefficient	Allowable coefficient
Zone C Terrain Transformation	Completion conditions	1.598	1.20
	Basic combination	2.721	1.20
	Earthquake combination	2.179	1.10

Table 6.7 Summary of earthwork engineering quantity (Unit: m^3)

	Zone A			Zone B		Zone C	
Project	Excavation	Filling	Slurry excavation	Excavation	Filling	Excavation	Filling
River channel	–	–	–	25,633	–	13,495	–
Terrain transformation	54,816	28,419	–	–	58,681	80,240	73,583
Outer embankment	–	7,847	–	–	3,545	–	191
Ecological filtering System	–	2,768	8,305	–	–	–	–
Total	54,816	37,902.4	–	25,633	62,226	93,735	73,774

6.3.2 *Earthwork Project*

The earthwork engineering quantity of the project mainly includes terrain transformation of newly-built wetland inland river and inner wetland, and the earthwork engineering quantities of the zones are shown in Table 6.7.

Chapter 7
Other Designs for Wetland Restoration Project

7.1 Construction Organization Design

7.1.1 Engineering Conditions

7.1.1.1 Project Profile

The project is Dianshan Lake Wetland Restoration Project Demonstration Zone in the western suburbs of Shanghai (625 mu), and is the Phase I project of the initial project of Dianshan Lake Wetland Restoration Project in the Western Suburbs of Shanghai. The project zone is located in Qingpu District of Shanghai, the Phase I project is located at the southeastern corner of the initial project of Dianshan Lake Wetland Restoration Project in the Western Suburbs of Shanghai, and takes up 625 mu totally.

The summary of major engineering quantity of the project is shown in Table 7.1.

7.1.1.2 Traffic Conditions

There are Huqingping Road, Lianxi Road and many municipal roads, and the landway transportation is convenient; Dalian Lake connects with Lanlu Harbor, Beiheng Harbor and Nanheng Harbor directly, and the water transportation is very convenient.

7.1.1.3 Supply of Water, Electricity, Labor and Materials

Electricity for construction assesses to the nearby electric network, the local telecommunication department can set up a communication line for construction communication or cell phones are prepared individually. Water for construction and

S. An and L. Wang, *Wetland Restoration: Shanghai Dalian Lake Project*,
Springer Environmental Science and Engineering, DOI 10.1007/978-3-642-54230-5_7,

Table 7.1 Summary of major engineering quantity

S/N	Project name	Unit	Engineering quantity	Remarks
1	Channel earthwork excavation	m^3	39,128	
2	Earthwork excavation of terrain modification	m^3	135,056	
3	Earthwork backfilling of terrain modification	m^3	160,683	
4	Outer embankment filling	m^3	11,583	
5	Slurry excavation of ecological filtering system	m^3	8,305	
6	Slurry backfilling of ecological filtering system	m^3	2,768	
7	Ecological bag	Piece	2,650	100 cm × 20 cm × 50 cm
8	Ecological grid	Piece	1,300	50 cm × 50 cm × 20 cm
9	Cinder	m^3	330	
10	Gravel	m^3	390	0.8–1.2 mm
11	Water percolation cloth	m^2	1,100	2–4 mm
12	Metasequoia	Tree	8,500	
13	Taxodium ascendens	Tree	5,000	
14	Taxodium Zhongshanshan	Tree	6,000	
15	Nymphaea tetragona	Tree	5,000	
16	Euryale ferox	Tree	5,000	

production can be collected from the nearby river channels. Domestic water is from the drinking water network.

7.1.1.4 Hydrological, Meteorological and Geological Conditions

The project is located in the south of the northern sub-tropic monsoon area, it has the classical marine climate, it is warm and wet and has four distinctive seasons, spring and fall are shorter, summer and winter are longer. Sunlight is sufficient and rainfall is abundant. The annual average temperature is 16 °C in the recent years, the highest temperature occurs in July and August, the average temperature is 28 °C in July, and the highest temperature is 38 °C extremely; the lowest temperature occurs in January and February. According to the meteorological information of Jinze Meteorological Station, the average evaporation is 1,300 mm, the annual average sunlight ratio is 1,500 h, the average humidity is 75 %, the average wind speed is 2.7 m/s, the frost-free period lasts 247 day, and there are abundant wetland and heat resources, which are suitable for growths of various crops.

The highest control water level of the river channel outside Jinze Town Grade 2 Dike is 3.5 m, the highest control water level of the river channel inside the dike is 3.2 m, the normal water level is 2.5 m, and the pre-floor precipitation is 2.0 m.

The underground water type in the area is the phreatic water stored in the surface soil, the supply resource comes from the atmospheric precipitation mainly and has the strong hydraulic connection with the river and lake waters near the site, and the static level buried depth of the groundwater is between 0.50 and 1.30 m (under the nature ground).

The soil layers under the 35.00 m of the natural ground in the site are the quaternary deposits, which are composed of cohesive soil, etc.

7.1.2 Drained Waters in Construction Period

The water drainage in the project construction period includes drained water of fish pool in the initial period and drained water in work space in the construction period. The project is located in the temperate subhumid monsoon area, and there are many rain days in dry season. The project shall be equipped with sufficient pumping and drainage devices to guarantee its planned implementation. It has five sets of JQB—2—10 submersible pumps.

7.1.3 Construction Methods

The project is separated into three areas, mainly are river excavation, topographical reform, filing and build peripheral bank and ecological filtering system, the earthwork will be carried out after drain water in fishpond, all construction will be carried out in dried earth.

7.1.3.1 River Excavation and Topographical Reform

The depth of river excavation is 1–2 m, excavation equipment is back digger of 1 m^3 hydraulic, and excavated soil will be carried into site of topographical reform with 10 t dump trucks. The soil can be dug, carried and back filled with back digger of 1 m^3 hydraulic for topographical reform, filing with the distance of within 500 m. And for those excess 500 m will be dug with back digger of 1 m^3 hydraulic, carried with 10 t dump trucks and level the site with 74 kW bulldozers.

7.1.3.2 Filling and Build Peripheral Bank

Fill and build peripheral bank with soil from area A, choose cohesive soil with better quality for it, separate the material with 74 kW bulldozers, damper with tractor, and make slop with back digger of 1 m^3 hydraulic. Compactedness of filling and build peripheral bank can't less than 92 %.

7.1.3.3 Ecological Filtering System

Excavation sludge for ecological filtering system will be dug out with back digger of 1 m^3 hydraulic, and dried in the sun and cured, then carried into appointed site with 10 t dump trucks, the fill and tamper. Fill and lay cinders, gravel by manual, lay permeable cloth, ecological bag, ecological case by manual.

7.1.4 Construction Layout

Construction layout is based on the principle of easy construction, saving land, saving investment, give attention to both the overall situation, focal important points. Give overall consideration and arranging reasonably for variety of permanent and temporary facilities, coordinate and interface properly works at every phase, ensure the work carried out smoothly.

7.1.4.1 Construction for Transportation

Personnel and equipment can enter through Liangxi road, or through Xicengzhi road, Lianhu road and get to bank of Dalian Lake. Dalian Lake, Beiheng port and Nangheng port are connect each other directly, water communication is very convenient, the cinders, gravel and other material needed for paving the road can be carried through water communication.

The road in the site will be arranged along with bank of lake, connect with variety working surface and temporary facility. Total length of road in the site is about 700 m, wide is 5 m, thickness of pavement is 20 cm, pavement is mud and stone combined.

7.1.4.2 Temporary Facility for Construction

Don't built special repair workshop for equipment and vehicle in order to save land, mechanical equipment can be repaired at repair workshop near by the construction site. The construction area is big, soil can placed in the site by coordination, don't need temporary site.

Primary machine equipment can be stored in parking area and canopy as temporary facility, the temporary facility is planned at B area where is relatively flat and filling is relatively less; it covered area is 300 m^2, it can store production material and tools, parking lots of machine equipment covers production material 300 m^2. Can rent civilian buildings for life and office facility, so only duty room of 100 m^2 is needed for construction at site. Detail refer to the "construction layout plan".

7.1.4.3 Power Supply, Water Supply

Power for construction can be connected from nearby power grid, water for construction can supplied from river or Dalian Lake, domestic water can be connected from tap water net.

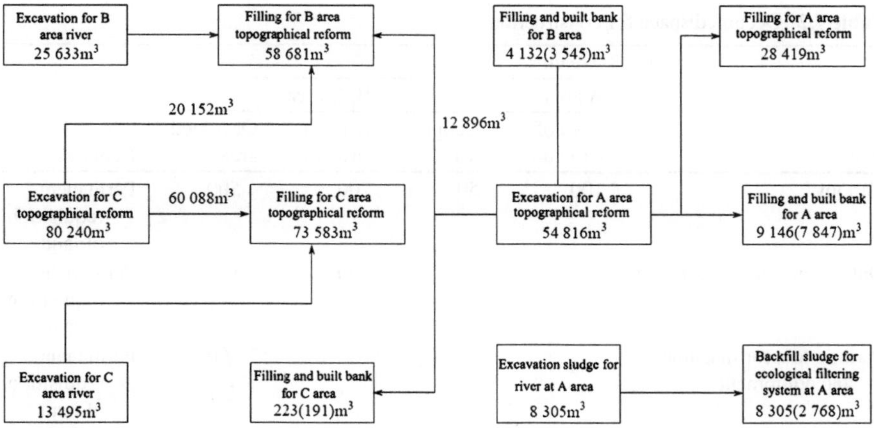

Fig. 7.1 Soil balance. Note: (1) Data in *bracket* is tampered volume, others are nature. (2) Nature soil: filling for topographical reform = 1:1, Nature soil: tampered for bank = 1:0.858, nature sludge: tampered = 3:1

7.1.5 Soil Balance

Volume of soil excavated from river is 39,128 m^3, volume of soil excavated for topographical reform is 135,056 m^3, and backfill for topographical reform is 8,305 m^3. Topographical reform of A area will be backfilled with excavated soil, and topographical reform of B area will be backfilled with excavated soil from river and excavated soil from topographical reform, shortage will be coordinated from A area. Soil for filling and built bund all are from topographical reform of A area. Detail refer to Fig. 7.1.

7.1.6 Construction Area

Area of temporary production facilities of the project covers 900 m^2. Duty room of the construction covers 150 m^2. Parking lots for machine equipment covers 2,500 m^2. Detail refer to the occupied area for Construction Table 7.2.

7.1.7 Construction Progress

Time limit for a project is 6 months. Preparation for the construction is 15 days, When all are available, dig A area and backfill, the duration is 3 months, and finish construction of ecological restoration system, finally carry out the afforest construction, the duration is 3 months. Excavation and backfill for B area will be carried out at third and fifth month, filling and built bank will be carried out at third and fourth month, the afforest construction of B area will be carried out at fifth and sixth month. Detail for variety progress refer to Table 7.3.

Table 7.2 Occupied space for construction

Item	Area/m^2				Remarks
	A area		B, C area		
	Area of structure	Occupied area	Area of structure	Occupied area	
Canopy	100	250	200	500	Permanent requisition of land
Duty room for construction			100	150	Permanent requisition of land
Parking lots for machine and equipment				2,500	Permanent requisition of land
Total	100	250	300	650	

Table 7.3 Construction strength summary

Item	Unit	Construction intensity index
Soil excavation	m^3/month	58,061
Backfill for topographical reform	m^3/month	53,561
Filling and built bank	m^3/month	5,792
Fill sludge for ecological filtering system	m^3/month	1,845
Permeable cloth	m^3/month	733
Cinders	m^3/month	220
Gravel	m^3/month	260
Ecological bag	Piece/month	1,767
Ecological case	Piece/month	867

7.1.8 The Main Machinery Equipment for the Construction

The required machinery equipment for the construction refer to Table 7.4.

7.1.9 The Main Technical and Economic Indexes for the Construction

The main technical and economic indexes for the construction refer to the Table 7.5.

7.2 Environment Protection Design

To strengthen the environment protection management of the construction project, in accordance with the related provisions of the *Environmental Protection Law of the People's Republic of China*, *the Regulations on the Administration of Environmental*

Table 7.4 Main machinery equipment for the construction

S/N	Construction machinery and equipment	Unit	Quantity
1	Back digger of 1 m^3 hydraulic	Set	9
2	74 kW bulldozers	Set	9
3	10 t dump truck	Set	36
4	74 kW tractor	Set	2
5	Scraper	Set	7
6	JQB—2—10 submersible pump	Set	5

Table 7.5 Main technical and economic indexes for the construction

S/N	Item	Unit	Indexes	Remarks
1	Total construction time	Month	6	
2	Labors at construction peak	Person	100	
3	Mean labors	Person	70	
4	Peak power consumption for construction	kW	30	
5	Peak water consumption for construction	m^3/day	120	
6	Temporary road	m	700	Road is 5 m, thickness is 20 cm, mud and stone combined

Protection of Construction Projects, *the Design Rules of Environmental Protection of Construction Projects*, engineering design documents shall contain the contents of environment protection design to implement environment protection measures.

7.2.1 Environment Protection Goals

7.2.1.1 Quality of Surface Water

The project is located in the Water Source Protection Zone of the upper reaches of Huangpu River. The waste water and domestic sewage produced in the course of the project construction shall be reused after treatment and strictly banned to be discharged into surrounding water bodies.

7.2.1.2 Quality of Ambient Air

The emission of air pollutants generated in the course of the project construction shall be controlled; air pollution source shall be controlled. The exhaust emission of the project shall reach the *Standard for the Integrated Emission of Air Pollutants* (GB16297—1996) class I, and the ambient air quality of the working area of the project and surrounding areas shall reach *Ambient Air Quality Standard* (GB3095—96) class I.

7.2.1.3 Quality of Sound Environment

The noise pollution produced in the course of the project construct shall be controlled. The noise in the construct site shall meet the related provisions of *Noise Limits for Construction Sites* (GB12523—90); the quality of the sound environment in the project area shall be in accordance with class I standard.

7.2.2 Environment Protection Standard

7.2.2.1 Environmental Quality Standard

1. *Environmental Quality Standard for Surface Water* (GB3838—2002) class II;
2. *Ambient Air Quality Standard* (GB3095—96) class I;
3. *Environmental Quality Standard for Noise* (GB3096—2008) class I.

7.2.2.2 Pollutant Emission Standard

1. *Integrated Emission Standard of Air Pollutants* (GB16297—1996) class I;
2. *Noise Limits for Construction Sites* (GB12523—90);
3. *Emission standard for Industrial Enterprises Noise at Boundary* (GB12348—2008) emission limits for class I functional area.

7.2.3 Analysis and Appraisal for the Affection of Project Construction on Environment

7.2.3.1 Analysis and Appraisal of the Affection of Project Construction on Environment

The Affection on Water Environment

The project is mainly wetland restoration and reconstruction project, basically without buildings. Its contents are mainly terrain transformation and ecosystem restoration and reconstruction. Small amount of production waste water, especially domestic sewage, will be generated in the course of construction. The number of constructors is about 100; water for life is 5 m^3/day; emission coefficient is 0.9; the maximum emission of domestic sewage is 4.5 m^3/day. The domestic sewage contains relatively many bacteria and pathogens, nitrogenous and phosphorous compounds and organic pollutants, etc. They will pollute water environment if they are directly discharged into rivers.

The water bodies in fishponds that contain higher organic content have serious breeding pollution and need to be drained quickly during construction. If any measure is not taken and they are let to be discharged into surrounding water bodies, they will cause serious organic pollution to the water bodies.

The Affection of Sound Environment

Excavators, tractors, trucks and other construction machinery and vehicles will be used in the course of implementing terrain transformation project. They will produce noise. It is estimated that the noise produced by them will be about 75–80 dB(A) and interrupt the normal life of surrounding residents for a short period of time.

Affection on Atmospherical Environment

The atmospherical pollution of the project will mainly come from the exhaust produced by oil-burning construction machinery and vehicles and the dust caused by the digging, reffing, transportation of the project. The exhaust and dust will fall the air quality of the construction site and surrounding areas and have adverse affection on the air environment in construction site and surrounding areas and sides of construction road for a short period of time.

Affection of Solid Waste on Environment

The earthwork digged within the project area will be used internally. The earthworks digged and refilled in the project are balanced, without need to buy. The loss of water and soil will be likely to be produced in the course of temporarily piling, transporting earthwork and have affection on surrounding soil and water environment. In addition, the likely falling of the earthwork and sludge in the course of transportation will also have certain affection on the environment along the transportation roads.

The constructors will produce certain domestic sewage during construction, its amount will about 50 kg/day. If it is not appropriately handled, it will pollute the soil and water within the project area and affect sanitation.

Affection on Ecosystem

According to investigation, the project will be mainly involved in water forests and fishponds. The water forest areas will not be interrupted, and the fishpond areas will be transformed and ecologically restored. The construction of the project will have less affection on ecological environment because the fishpond areas are greatly affected by human affection and basically have no aquatic plants.

Affection on Human Health

The constructors of the project are not many, but the hygienic condition on the construction site is relatively weak, which provides spread opportunity to infectious diseases.

Affection on Traffic

The transportation vehicles of the project will increase local traffic pressure.

7.2.3.2 Analysis, Prediction and Appraisal of Operational Affection

Affection on Water Environment

After its implementation, the project will reduce the breeding pollution around Dalian Lake and fall the pollutants discharged into water bodies. At the same time, the pollutants in the boiled water bodies can be decreased further by forest wetland and shoal wetland which have strong pollution-purifying ability. Besides, as water bodies increase in volume, water environment capacity will boost accordingly. This will improve the quality of the water bodies of Dalian Lake and surrounding areas to certain extent.

Affection on Biological Resources

1. Affection on the biological resources of wetland.

   The arbors bearing colorful flowers and aquatic plants will be planted in accordance with the ecological needs and landscape requirements of different areas in the course of the project construction. Through reasonable configuration of the arbors, the biodiversity in the project area will be conspicuously improved and good living environment will be provided to birds, frogs and other small animals. Thus, the stability of the ecosystem in the project area will be promoted, and good lake ecosystem will be formed.

2. Affection on birds resources.

   After the restoration of shoal wetland, low grass, small shrubs, moderate amount of arbors and other plants will be planted to create the ecosystem suitable for birds to perch and live, so that birds will come here to find food and perch. At the same time, the improvement of the water quality in lake area will increase the food of birds and improve the living environment of birds. Therefore the types and amount of birds will go up to certain extent after the implementation of the project and the stable and good development of ecosystem will be promoted.

Hence, the implementation of the project will play a very positive role in the restoration and development of the ecological system in Dalian Lake and surrounding areas.

Affection on Landscape

The project will help improve and enrich the landscape of the project area, increase wetland landscape effect and enhance the aesthetic value of landscape by transforming the existing fishponds into relatively typical forest wetland and shoal wetland and by expanding the existing water forests.

Affection of Digging, Crushing and Moving

The land circulation in the project area has been carried out, and the life and production of the residents will not be affected.

7.2.4 Environment Protection Design

Targeted at the adverse affection probably produced during project construction and operation, corresponding measures shall be taken to reduce and avoid it through environment protection design.

7.2.4.1 Aquatic Environment Protection Design

The project will not generate waste water basically; the maximum emission of domestic sewage will be about 4.5 m^3/day. When temporary working sheds are set up on construction site, transient toilets and storage pools of $2\ m \times 1.5\ m \times 1.5\ m$ shall be established to collect domestic sewage and periodically transport it.

The following management measures shall be taken to further reduce and avoid the affection of the project on surrounding water environment during construction and operation.

1. Attention shall be paid to cleaning construction site and timely repairing construction machinery to prevent oil from leak. If any leak, measures shall be taken in time; special devices shall be used to collect and appropriately dispose it.
2. Efforts shall be made to strengthening the education of the constructors to carry out the principle of civilized construction and strictly follow operational regulations to avoid and reduce pollution accident.
3. The earthwork shall be piled on the appointed places, effective measures shall be taken to protect it from rain, and timely refilling shall be performed to prevent it from running into rivers with surface runoff.

4. Supervision and management of construction site shall be strengthened; the domestic sewage and untreated construction production waste water are banned to be discharged into rivers and Dalian Lake.

7.2.4.2 Acoustic Environment Protection Measures

The requirements of construction noise of *Noise Limits for Construction Site* (GB12523—90) shall be strictly exercised. The following protective measures shall be taken to reduce the affection of construction noise as far as possible:

1. Low noise construction machinery shall be adopted as far as possible and kept to be in good condition.
2. The construction vehicles shall not be run on the roads near to the places where people are concentrated as possible as they can.
3. The construction unit shall rationally arrange construction time; high noise equipments shall be banned to operate between 22:00 and 6:00 next day.
4. The high noise equipments shall be so reasonably allocated to make them far from resident areas.
5. The construction unit shall intensify construction management, conduct civilized construction and reduce unnecessary artificial noise.
6. Constructors shall wear earmuffs and sound proof helmets when they work in strong noisy environment; when noise is over 85 dB(A) and there are not protective measures, contact time shall be decreased pursuant to related provisions.

7.2.4.3 Air Environment Protection Measures

The construction unit shall use the construction machines with good property as possible as it can. The machines that do not conform to national exhaust emission standards shall be banned to enter the construction area, and the emission of pollutants and the exhaust that exceeds standard shall be strictly controlled. At the same time, construction vehicles shall slow down in construction area. The earthen roads in the construction site and the roads on which construction vehicles frequently run shall be often cleaned and sprayed with water to keep clean and certain humidity to avoid raising much dust.

7.2.4.4 Measures of Solid Waste Affection

The earthwork digged in the project shall be timely refilled or transported away. The earthwork digged on that day shall be used on the same day. The earthwork that can not be used on the same day shall be covered with felted cloth and other dust proof and rain proof measures. Measure shall be taken to avoid the earthwork falling from vehicles in transportation.

The existing sanitary facilities shall be used to collect the life garbage of constructors. In the places where the existing sanitary facilities can not be used, the management of the sewage life garbage shall be strengthened, uniform garbage cans shall be set up by section and classification, and sanitation departments shall be entrusted to clean and transport the garbage periodically. Randomly discarding garbage shall be banned.

7.2.4.5 Sanitation Management in Construction Area

Attention shall be paid to sanitation and epidemic prevention in construction area during construction period; the following measures shall be taken:

1. Constructors shall take medical examination in medical and health care institutions before they enter construction area. The constructors who are sick can not enter the construction area until they are cured.
2. For constructors' health, the health and epidemic prevention publicity and education in the construction area shall be strengthened; the work of health and epidemic prevention shall be done well.
3. Sanitation management system in construction area shall be formulated; the examination of the sanitary condition in the construction area shall be intensified.
4. Dietary hygiene management shall be strengthened to avoid dirty food and drinking water so that the breakout and epidemic of hepatitis and dysentery can be prevented.

7.2.5 Environmental Management and Monitoring Plan

7.2.5.1 Environmental Management

The environmental management of the project during construction period shall be carried out jointly by the owner unit, construction unit and supervision unit. Their main duties shall include the coordination and unified management of environmental protection work during construction period, implementation of the environmental protection measures from beginning to the end of the project construction, coordination with local environment protection department to do well the work of the environment monitoring and examination of the project. The construction unit shall be responsible for implementing environmental protection measures during construction period, and the owner unit and supervision unit shall be responsible for the supervision and management of the implementation of the environmental protection measures, quality and progression, etc.

The Bureau of Water Affairs of Qingpu District shall take environmental management duty during operation period and arrange professionals to conduct uniform environment protection management.

7.2.5.2 Environmental Monitoring Plan

During construction period, environment monitoring departments with qualification can be entrusted to carry out environment monitoring tasks.

Water Environment Monitoring

1. Monitoring station.
   One monitoring station shall be established in Dalian Lake area to monitor the water quality of the lake; one monitoring station shall be set up at the river within project area.
2. Monitoring index.
   The monitoring indexes include water temperature, pH, DO, BOD, COD_{Mn}, NH_3-N, total phosphor, total nitrogen, petroleum and total suspended solids.
3. Monitoring frequency.
   The monitoring frequency of the quality of surface water: once per season in construction period and once per half year in early operation period (1 year after construction period).

Acoustic Environment Monitoring

1. Monitoring station.
   Two mobile monitoring stations shall be set up at the places near to construction area and relative sensitive in accordance with the construction progress of the project.
2. Monitoring item.
   Equivalent consecutive class A sound.
3. Monitoring frequency.
   The monitoring time is construction period, monitoring frequency is once every 2 months, twice altogether. The noise monitoring is divided into two time periods, am 6:00–pm 10:00 and pm 10:00–am 6:00, sampling for 15 min respectively.

7.3 Water and Soil Conservation Design

7.3.1 Design Basis

1. *The Water and Soil Conservation Law of People's Republic of China* (June 1991);
2. *The Implementation Regulations on the Water and Soil Conservation Law of People's Republic of China* (Order No. 120 of State Council 1993);
3. *Administrative measures for the Water and Soil Conservation Plan of Development and Construction Projects* (Ministry of Water Resources,

State Development Plan Commission, State Bureau of Environment Protection Water Resources Protection (1994) No. 513);

4. *Technical Specifications for the Water and Soil Conservation Plan of Development and Construction Projects* (SL204—98);
5. *Soil Erosion Classification and Grading Standard* (SL190—96);
6. *Technical Specifications for Water and Soil Conservation and Monitoring* (SL277—2002);
7. *Charging Standard on Engineering Investigation and Design* (National Development Plan Commission, Ministry of Construction (2002) No. 10, document);
8. *Budget Estimation Provisions of Water and Soil Conservation Projects* (water total (2003) No. 67, document);
9. *Announcement on Dividing Key Prevention and Control Areas of National Water and Soil Loss* (April 2006, Announcement of Ministry of Water Resources, No.2, 2006);
10. *Guidance Opinion on the Consultation Service Cost of the Water and Soil Conservation of Development and Construction Projects* (Conservation Monitoring (2005) No. 22);
11. Other related technical data in which the project is involved.

7.3.2 Fundamental Principle

1. *The Water and Soil Conservation Law of People's Republic of China* shall be exercised in an all-round way; the policy of "Prevention First, All-round Planning, Comprehensive Prevention and Control, Right Measure for Right Land, Strengthening management, paying attention to profit" of water and soil conservation shall be adhered to.
2. The integration of engineering measures and plant measure shall be stuck to. Measures for forests and grass shall be taken to the hilt under the prerequisite of guaranteeing the safety and standard of water and soil conservation measure so as to increase green land area and beautify environment.
3. The adaption of water and soil conservation measures and major project construction shall be kept to. The water and soil conservation measures shall be made to be technically feasible and economically reasonable in accordance with the requirement of "construction, design and acceptance at the same time" and adhering to scientific and economic principle.

7.3.3 The Status Quo of Water and Soil Loss

The project is located at Jinxe Town, Qingpu District, Shanghai, belonging to plain with river network. The place where the project is belongs to key monitoring area of water and soil loss according to the *division method of the water and soil loss of*

Jiangsu Province near to the project, and allowed value of water and soil loss is 500 t/(km^2 · a).

The water and soil loss types of the project area are mainly rainfall surface erosion and water erosion caused by surface runoff and the water and soil loss resulted from human activities of development. The main representation forms of the water and soil loss of the project area are slope surface erosion, ephemeral gully erosion, etc. According to the surface erosion grading index form of *Soil Erosion Classification and Grading Standard* (SL190—96), the project area belongs slight erosion area, with soil erosion modulus background value 400 t/(km^2 · a).

7.3.4 Earthwork Balance

The digged earthwork of the project is 174.2 thousand m^3, refilled earthwork is 174.2 thousand m^3, digged sludge earthwork is 8.3 thousand m^3. All of them shall be used for quick infiltration system.

7.3.5 The Responsibility Scope and Subarea of Water and Soil Loss Prevention and Control

7.3.5.1 The Responsibility Scope of Water and Soil Loss Prevention and Control

According to *Technical Specifications for the Water and Soil Conservation Plan of Development and Construction Projects* and the principle of "whoever develops, whoever protects; whoever causes water and soil loss, whoever is responsible for control", the responsibility scope of water and soil loss prevention and control is determined to contain project construction area and direct affection area.

The project construction area: the total area of the project construction area is 625 mu. Temporary facilities of construction occupy an area of 1.35 mu (located in the project area).

The direct affection area: refers to the area which will have water and soil loss affection because of construction activity affection. The project and construction site account for 15 mu altogether outside the project area.

7.3.5.2 Subarea of Water and Soil Loss Prevention and Control

The project construction area and direct affection area of the responsibility scope of the water and soil loss prevention and control of the project are divided into two subareas of water and soil loss prevention and control through primary analysis, in accordance with the terrain of the project, construction plane layout, the functions

Table 7.6 The statistic form of the subareas of prevention and control of the project and total area

The subareas of prevention and control	Project construction area/mu	Directly-affected area/mu	The responsibility scope of prevention and control/mu
The subarea of the prevention and control of the main part of the project (I)	623.5	15	638.5
The subarea of the prevention and control of temporary facilities (II)	1.35	0	1.35
Total	625	15	640

of different places, the characteristics of water and soil loss, proposed different measures of water and soil conservation. They are:

1. The subarea of the prevention and control of the main part of the project (I).

The subarea of the prevention and control of the main part of the project occupies an area of 623.65 mu (excluding the area occupied by temporary facilities), including the land occupied by the main part of the project and the area directly-affected by it.

2. The subarea of the prevention and control of temporary facilities (II).

It includes the scope occupied by production and life facilities and relevant affected scope, 1.35 mu altogether.

The subareas of prevention and control of the project and total area are seen in Table 7.6.

7.3.6 The Analysis of the Affection of Development and Construction Activities on Water and Soil Loss

The new water and soil loss will be from the following aspects during construction: (1) the vegetation or facilities around Dalian Lake that have the function of water and soil conservation will be affected to certain extent, which will reduce and even lose their function of water and soil conservation; (2) The contemporary construction facilities will occupy some land and press, bury or damage existing vegetation, causing water and soil loss.

7.3.7 The Goal of Water and Soil Loss Prevention and Control

In accordance with the related provisions of the *Technical Specifications for the Water and Soil Conservation Plan of Development and Construction Projects* (SL 204—98), the measures of water and soil conservation shall be taken to prevent

Table 7.7 The specific goals of water and soil loss prevention and control

S/N	Item	Goal of prevention and control (%)
1	Interrupted land control rate	95
2	Water and soil loss control rate	85
3	Water and soil control proportion	1.2
4	Dreg stopping rate	95
5	Grass, forest and vegetation restoration rate	95
6	Forest and grass coverage	20

and control the new water and soil loss caused by project construction and production. Water and soil resources shall be protected, improved and reasonably used under the premise of guaranteeing safety operation to heighten land productivity and rebuild good ecological environment. The project area that is located at the plain with river network belongs to the key monitoring area of water and soil loss. In accordance with the area of water and soil loss prevention and control where the project is, regional ecological function of water, soil conservation and the project reality, the project is a construction project and will implement class II standard of water and soil loss. The specific goals of water and soil loss prevention and control are seen in Table 7.7.

7.3.8 Measures of Water and Soil Loss Prevention and Control

7.3.8.1 The Measures and Layout Principle of Water and Soil Conservation

The responsibility scope of prevention and control is divided into four subareas of water and soil prevention and control in accordance with the functions of different places of the project and the characteristics of water and soil loss. Reasonable measures of water and soil conservation shall be taken in each subarea in accordance with the characteristics of the development and construction of the main part of the project and based scientific basis of water and soil loss prediction so as to prevent and control new water and soil loss and improve the ecological environment of the project area. At the same time, the measures of plants shall be taken pursuant to the specific conditions of each subarea and associated with the engineering measures of water and soil conservation that has been taken in order to increase vegetation coverage and lessen surface runoff. The integration of projection development, prevention and control and combination of "points, lines and plane" will be conducted to complete system of water and soil loss prevention and control.

Table 7.8 The measure system of water and soil conservation

Subarea	Measures of water and soil conservation of the main part of the project that has been taken	New measures of water and soil conservation
The subarea of prevention and control of the main part of the project (I)	Wetland restoration with respectively strong function of water and soil conservation	Temporary drainage measure of construction, tilling and planting soil coverage
The subarea of prevention and control of temporary construction facilities (II)		Temporary stopping measure of construction, sand subsiding measure, slope leveling after construction

7.3.8.2 The Measure System of Water and Soil Conservation

The requirements and proposal of water and soil conservation shall be put forward targeted at the links that cause water and soil loss in project design and construction design, based on the analysis and appraisal of the overall layout, construction organizing, construction workmanship and spoil treatment mode of the main part of the project and the protective measures of water and soil conservation and associated with the division and reasonable all-round and systematic plan of the subareas of prevention and control. Some new measures of water and soil conservation of all kinds of engineerings shall be taken to form a complete system of water and soil loss prevention and control that is guided by project measures and associated with land control and plants (Table 7.8).

7.3.8.3 The Measure Design of Prevention and Control of Subareas

The measures of water and soil loss prevention and control of each subarea.

1. The subarea of the prevention and control of the main part of the project (I).

The subarea of the prevention and control of the main part of the project occupies an area of 623.65 mu. The main measure of water and soil prevention conservation is the establishment of temporary drainage ditches and sand subsiding ponds at the two sides of temporary earthwork piling grounds and coverage of temporary earthwork piling grounds.

Main project workload: drainage ditches of 2,080.8 $m^3$and 16 sand piling ponds.

2. The subarea of the prevention and control of temporary facilities (II).

It includes the scope occupied by production and life facilities and relevant affected scope, 1.35 mu altogether. After construction is finished, above ground buildings and hardened ground shall be removed completely, construction garbage shall be cleaned, ground shall be leveled, slash shall be restored and grass seeds shall be sown.

Table 7.9 New measures and project workload of water and soil loss prevention and control

Subarea	Measure	Unit	Quantity
The subarea of prevention and control of the main part of the project (I)	Drainage ditches digging	m^3	100
	Earthwork covering	m^2	50
The subarea of prevention and control of temporary construction facilities (II)	Ground leveling	mu	1.35
	Grass seeds	mu	1.35

Main project workload: ground leveling of 1.35 mu, grass seeds sowing of 1.35 mu.

According to the measures of water and soil conservation of each subarea, new measures of water and soil loss prevention and control are seen in Table 7.9.

7.3.9 Water and Soil Loss Monitoring

Comprehensive means shall be taken to monitor the reasons, quality, intensity, affection scope and consequence of water and soil loss starting from the angle of protecting water and soil resources and associated with the construction and management of the project and characteristics of local water and soil loss so as to understand the harm caused by the water and soil loss of the project, put forward corresponding measures of prevention and control. The implementation and protective effect of the measures of water and soil conservation shall be monitored and appraised to take full advantage of monitoring results to guide the work of water and soil loss prevention and control while provide important basis for the completion and acceptance of the project.

Monitoring contents include the amount of water and soil loss, the effect of the measures of water and soil conservation. In accordance with the characteristics of the loss and prevention and control of water and soil, the method of integrating fixed point observation, investigation and monitoring shall be adopted to monitor water and soil conservation. Fixed location observation shall be given more attention than field investigation.

7.4 Project Management and Maintenance

7.4.1 Management Institution and Personnel Composition

After the first stage project of the wetland ecosystem restoration of Dalian Lake, the government of Qingpu District, Shanghai, will organize special institution for uniform management. The government will also organize, coordinate and supervise its related managements to conduct long term coordinated management associated

Table 7.10 Project managements list

S/N	Position	Number
1	Unit principal and administrative management	1
2	Technical management	1
3	Operator and repairer	3

with the wetland of surrounding areas, so that maximum benefit of the project will be exerted.

The project is newly-built wetland, occupying an area of 625 mu. The managements of ecological restoration project mainly consist of unit principals, administrative managements, technical managements, financial and asset managements, water supervisors, operators, observers, repairers and auxiliaries, etc. The managements of other engineerings will be determined in accordance with need.

Management institution shall reduce structural layers and non-production personnel as possible as it can by carrying out the principle of simplification and high efficiency and reasonably setting up functional departments or positions. The institution will increase five managements in accordance with the requirements of project operation and management. The specific positions are seen in Table 7.10.

7.4.2 Management Scope and Contents

7.4.2.1 Management Scope

The management scope of the project includes rivers, ecological islands, forest, shoal greening and organic vegetables and breeding in the project area.

The protective scope of the project refers to certain areas divided outside the management scope of the project to protect the safety of the project, but it is not taken as collecting land scope. Activities that harm project safety, such as, hole-digging, well-sinking, explosion are not allowed in the area. If necessary, they shall be approved by related authorities. The protective scope of the project is 50 m outside the management scope.

7.4.2.2 Management Contents

The management contents include the management of dikes, slope protection, water surface, aquatic and terrestrial vegetation, etc., and water quality monitoring and management. Systematic plant protection shall be conducted to promote the benign development of the ecological environment of water bodies, fully exert the benefit of ecological restoration project and maintain good aquatic ecological environment.

7.4.3 *Water Quality Monitoring and Biological Investigation*

The water quality monitoring and biological investigation shall be carried out during operation period to master the implementation effect of the project and understand the restoration state of ecological system and the improvement condition of water quality after the project construction is ended.

7.4.3.1 Water Quality Monitoring

1. Monitoring station: One water quality monitoring station shall be established at A area, B area, C area and Jinmengkou Port respectively, 4 monitoring stations altogether.
2. Monitoring frequency: Once per season, four times per year.
3. Monitoring index: Water temperature, pH, DO, BOD_5, COD_{Mn}, NH_3-N, total phosphor, total nitrogen, petroleum and total suspended solids, ten items altogether.

7.4.3.2 Biological Investigation

1. Investigation location: One investigation location shall be set up at A area, B area, C area respectively. More can be established in accordance with plant growth state if necessary.
2. Investigation time and frequency: The biological investigation shall not be conducted within 1 year after the project construction is finished to reduce interruption in the area of project and give certain time and space of development to the ecological system within the project area. The first biological investigation shall be carried out after 1 year, once in April which is plant germinating period and once in August which is plant maturing period. After the first investigation, the biological investigation shall be conducted once or twice a year until the ecological environment is relatively stable.
3. Investigation contents: Types and amount of terrestrial and aquatic plants, terrestrial and aquatic animals, birds, planktonic algae and zoobenthos as well as the growth and subsistence of invasive species.

7.4.4 *Annual Operation and Management Cost*

The annual operation and management cost of the project shall be primarily estimated for the reference of the decisions of departments by reference to the statistic data of the annual operation and management cost of similar projects of related areas that have been finished so as to guarantee that the project can normally work and achieve anticipated benefit. The annual operation and management cost includes project maintenance cost, salary and welfare, material cost, other direct costs, management cost, monitoring cost and investigation cost, etc. It is about RMB 400 thousand yuan per year during operation period through estimation.

7.4.5 Project Operation and Management

The management institution shall enact specific management system and operation rules in accordance with *Water Law, Shanghai Greening Management Regulations, the Implementation Guidelines of the Protective Regulations of the Water Source of the Upper Reaches of Huangpu River in Shanghai* and other central and local laws and regulations.

1. Maintenance of wetland system: Scientific operation and management not only can maintain the stable and high efficient effect of wetland in removing pollutants but also can gradually decrease human interruption and realize the automatic operation of ecological system.
2. Planting: The planting period is usually winter and after frost or autumn before frost. 4–5 weeks of cultivating period shall be guaranteed for the plants that have been planted before water enters wetlands.
3. Harvest management: For wetland, conventional plant management is unnecessary, because plant community has good self-maintenance property. Plants grow and die and continue to grow next year. Under favorable environmental condition, they can naturally spread to unsown places and also can move from the places under relatively greater environment pressure. The managements can control the scope of the spread of plants through harvest.

As for a healthy wetland ecosystem, harvesting plant is not necessary. Removing died plants can make new plants grow stronger next spring. Burning plants in winter can control pests, maintaining some fallen leaves can increase the insulation of sand and gravel surface to make wetland system maintain relatively high temperature. But from aesthetic angle, harvesting plants every autumn makes plants grow stronger and more aesthetic next spring.

The aquatic plants of the project are of both ecological value and certain economic value. Therefore, they shall be harvested in maturing period, so shall the organic fish of the project. It is necessary to pay attention to two principles in harvest. One is timely principle; the other is moderate amount principle. For the timely principle, both the life history of plants and animals and economic benefit shall be taken into account; moderate amount principle refers to harvesting strictly in accordance with logistic growth curve to retain the self-production ability of organism.

7.4.6 *Planting and Management Methods of Wetland Plants*

The following is the common planting and management methods of wetland plants.

7.4.6.1 The Multiplying and Planting Technology of Lotus

1. Planting time: Generally speaking, lotus is planted when local daily average temperature is above 15 °C, and water temperature over 12 °C, at the end of March or in the middle of April in Yangtze River basin, in advance accordingly in south China and later accordingly in north China.
2. Reasonable close planting: Lotus shall be reasonably closely planted with plant line distance 0.7 m×1.5 m; 400–500 plants/hm^2, seed amount 150–250 kg (seedling lotus 120–200 plants).
3. Planting approach: The experimental results of Wuhan Plant Research Institute and other units indicate that it is suitable to adopt horizontal planting method in order to make lotus roots germinate and grow in advance.

7.4.6.2 The Multiplying and Planting Technology of Gorgon Fruit

1. Water level adjustment: Water is suitable in 30 cm depth when gorgon fruit is planted, increase to 40–60 cm after it is alive, increase further to 1 m at late strongly growing period and early flowering and bearing fruit period, but it is not appropriate to rise water level up to 1.2 m in a short period of time, slowly fall to 50 cm at late flowing and bearing fruit period. In short, water level is adjusted in accordance with the growth of gorgon fruit.
2. Examining and complementing seedling: Examine seedling 10 days after planting; if lack, complement it in time. The seedling that grow weakly shall check whether its core leaf is covered by sludge or not. If yes, the sludge shall be removed in time, or it shall be pulled out and replaced with new seedling. Good harvest can be obtained as long as complete and strong seedling is guaranteed because gorgon fruit is planted with low density.
3. Weeding: Weeding shall be conducted for 2–4 times in accordance with the state of weed in field before gorgon fruit grows to cover lines in field. The removed weed can be stepped into mud as fertilizer. Associated with weeding, heaping mud over gorgon fruit roots shall be performed to protect roots until pits are level. Don't damage gorgon fruit leaves and roots. Especially in operation, core leaves can not be buried by sludge.

7.4.6.3 The Multiplying and Planting Technology of Water Chestnut

1. Water level adjustment: As water chestnut grows, water level shall be gradually deepened to maintain plants grow strongly. Water level shall be adjusted dependent on water chestnut varieties. It is a taboo to submerge water chestnut plants into water.

2. Examining and complementing seedling: Examining seedling shall be exercised everyday after water chestnut is planted. If lack of seedling and floating seedling are found, complementing seedling shall be carried out in time.
3. Weeding: Weeding shall be conducted for 3–4 times in accordance with the grow of weed in field before water chestnut grows to cover lines in the field. Commonly-seen weed is Nymphoides, Salvinia natans, eel grass, pondweed, etc.

7.4.6.4 The Multiplying and Planting Technology of Zizania aquatica

1. Attaching parent plants to zizania aquatica plants to grow seedling: It is the important improvement in the planting technology of the zizania aquatica of two crops a year in recent years. Chosen parent plants first are digged out in autumn, then attached to zizania aquatica plants for sometime, later parted with zizania aquatica plants and planted to big field. This method of growing seedling can reduce using time of big field, helpful for crop rotation and improve zizania aquatica mature in advance and reduce damage of zizania aquatica seedling in summer while greatly improve the purity of the zizania aquatica seedling in autumn next year. Attaching seedling shall be performed the level land near planting field and fertilized with 7,500 kg/hm^2 organic fertilizer as basic fertilizer. Too much fertilizer is not suitable to avoid overgrowing. Attaching seedling shall be carried out during December–January next year.
2. Planting in spring: It is suitable for most varieties of the zizania aquatica of two-crops a year and all varieties of the zizania aquatica of one crop a year, and planting time is late April in the middle and lower reaches of Yangtze River. zizania aquatica shall be planted when even soil temperature is above 15 °C and when seedling grows to about 20 cm. The parent plants shall be digged out with mud from seed field. A sharp knife shall be used to vertically cut tillers into small pieces without damage the new roots of the tillers. Each piece shall have old stem and healthy 3–5 tillers, planting while digging. If introducing seedling, the seedling must be kept humid in transportation. Planted zizania aquatica seedling plants can be cut short if they are too high. Their suitable height is about 30 cm to reduce water vaporing and swinging after planting. The line distance of planted plants depends on planting mode, variety, tillers of each piece, planting time, soil fertility and other factors. Generally speaking, line distance is 60–100 cm, plant distance is 60 cm, 16,000–20,000 pits/hm^2, about 90,000 plants.

7.4.6.5 The Multiplying and Planting Technology of Cattails

1. Planting: Cattails seedling plants are chosen and planted when temperature go up to 15–20 °C and after they sprout. Planting time is mostly in April or May. Strong plants are chosen from parent cattails field as seedling plants. The seedling plants are digged with more roots and mud while they are transported and planted. Some leaves shall be cut if they are too long to avoid vibration after

planting. The suitable line distance and plant distance of planted plants is 50 cm × 50 cm, planting depth is 10–15 cm to make roots enter into mud against falling or floating.

2. Field management: Water depth 10–15 cm shall be maintained after planting to avoid drought which restrain growth and arouse big amount of heading and flowering. Water level shall be deepened as the plants grow, generally 60–80 cm, maximum 100 cm. If water level is too shallow, cauloid will be short with low quality; if too deep, the cauloid will be long and thin, and quality and yield will be affected.

7.4.6.6 The Multiplying and Planting Technology of Reed

Planting reeds is the work of moving reeds from one place to another whether they are bought from seedling nursery or digged from lakes or transferred from big seedling plants. It is necessary to achieve production purpose through long term human management.

1. Line distance and plant distance: Line distance and plant distance are 30–50 cm for the reed seedling plants that are bred through seed, 60–120 cm for the reed seedling plants that are transferred from reed field.
2. Planting depth: 5–15 cm.
3. Planting time: At the end of March.
4. Water level adjustment: Reed has different requirements of water level in different growth periods. The principle first from shallowness to deepness and then from deepness and shallowness shall be followed in water level adjustment. At the same time, the water level shall be stable. Generally speaking, water level is 5–10 cm when the reed is planted, increase to 20–40 cm as the reed grows, fall to 5 cm when the reed heads, field drying after the reed completely maters to facilitate mechanical harvest.

7.4.7 Management Laws and Regulations

Wetland restoration project shall be managed in accordance with related national, local and industrial laws and regulations, such as the *Flood Control Law of the People's Republic of China, Shanghai River Management Regulations, the Flood Control Regulations of the People's Republic of China, Management and Design Specifications for Dike Projects (SL171—96), Shanghai Greening Management Regulations, the Implementation Guidelines of the Protective Regulations of the Water Source of the Upper Reaches of Huangpu River in Shanghai, The Water Pollution Prevention and Control Regulations of the People's Republic of China*, etc.

Chapter 8
Construction and Management of the Project

8.1 Construction Overview of the Project

8.1.1 Basic State of the Project

The project is the Wetland Restoration and Reconstruction Area (Area A) of Shanghai Xijiao Dianshan Lake Wetland Restoration Model Area Project (625 mu), belonging to first stage construction model area, with total area 150 mu. The project is divided into three stages in implementation. The first stage project is terrain moulding and biodiversity improvement projects. Its main contents include accumulated water drainage, sludge digging, aeration, islands slope filling and construction, wetland bottom construction, etc. The second stage project is micro-topography wetland moulding and quick infiltration system project, mainly including the terrain transformation of deep water area, quick infiltration system construction, etc. The third stage project is the restoration system project of forests, shrub wetland and coast wetland, mainly including the vegetation plant of islands and shore slope belt, and the vegetation restoration of waters, etc.

8.1.2 Construction Characteristics of the Project

1. Mechanical construction is very difficult. Because the first stage project and the second stage project are all involved in earthwork digging, sludge amount is great and needs to be aerated, it is necessary to dig repeatedly for many times, restricting construction period. Only mechanically-operated small dumpers can be used for construction.

S. An and L. Wang, *Wetland Restoration: Shanghai Dalian Lake Project*, Springer Environmental Science and Engineering, DOI 10.1007/978-3-642-54230-5_8,

2. It is necessary to take the requirement of scientific research into account. Because it has scientific purpose, the project has relatively higher requirements in the moulding of wetland terrain and vegetation planting. Therefore, the requirement of scientific research must be taken into consideration, including the distribution of water treatment system.
3. Workload is huge, work is relatively concentrated. Working procedure must be arranged in advance to avoid the interruption of intercrossing tasks.
4. Capital lag shall be coordinated with work delay of national season, allocating workforce, machinery and capital must be made in advance to ensure that the project shall be progressed in accordance with the plan.

8.1.3 Overall Construction Layout

On-site plane layout shall be made in accordance with on-site observation and associated with the actual requirement of the project. Near houses of residents or units shall be rented as life facilities, offices, materials and equipments warehouses.

8.2 Construction Organization and On-site Management

8.2.1 Construction Organization Structure

The construction organization structure of the project is seen as Fig. 8.1.

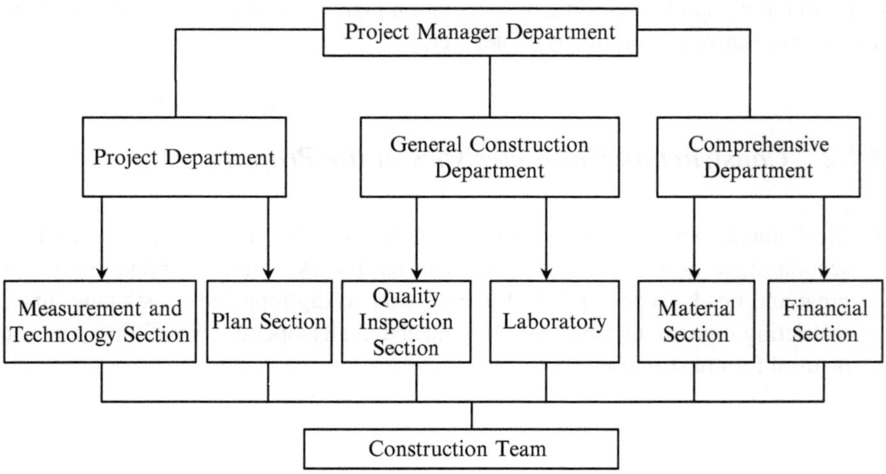

Fig. 8.1 Construction organization structure

Project manager department led by project manager shall be established to implement construction management of project and exercise the functions of plan, organizing, coordination and supervision and conduct uniform plan, management and scheduling. The project manager department shall establish business departments responsible for the normal work of each business of the project and connection with higher authorities, proprietor and supervisors. Construction teams shall be mainly responsible for specific construction on site. Each institution shall have fixed personnel, fixed posts and fixed responsibilities, strictly carry out contract terms and finish all kinds of management work with high quality and high efficiency.

The work and management responsibilities of main personnel and institutions:

1. Project manager: Presides work in an all-round way, exercises project contracts in an all-round way, responsible for project quality, safety, construction period and cost control, and conducts internal administrative management of the project manager department, including personnel dispatch, financial management and foreign coordination, etc.
2. Vice project manager: Mainly takes charge of safe production, civilized construction, construction progress, resources allocation, team management, responsible for organizing and directing on-site production, interfacing, coordinating and internally examining production teams.
3. Chief engineer: Mainly takes charge of technical work and quality control, coordinates with supervision units, design units, quality inspection stations and proprietor and manage technical department of construction and quality inspection department.
4. Technical department of construction (with measurement team and laboratory): Organizes, designs and reviews documents, manages construction drawings, contracts and technical specifications in an all-round way. It organizes and designs construction in accordance with the requirements of contracts and takes charge of the measurement and experiment of the project and coordinates with design department and supervisors. In addition, it files and sorts out designs, modifications, claim documents and negotiations, sets up technologies and management diaries to do well technical archive management of the project. It knows well project progress of each production unit, files and analyzes the factors affected progress and proposes improvement measures. It organizes key technology research, takes charge of technical disclosure, inspects and guides the technical work of construction teams.
5. Quality department: Formulates the management regulations and guarantee measures of technologies and quality in line with the specific condition of the project and associated with the management characteristics of the project; proceeds quality activities in an all-round way pursuant to quality system; takes charge of the inspection, appraisal, acceptance, quality management of the project and coordinates with design and supervision departments.
6. Quality department: Responsible for on-site safety; enacts safety regulations and systems, inspects the implementation of safety systems to guarantee exercise of the safety measures of construction; periodically holds safety knowledge education and the training of special work types.

7. Financial department: Does well the work of capital use and management; takes charge of financial statements analysis; analyzes and predicts expenditure and seeks for the possibilities of cost saving.
8. Material section: In accordance with the construction drawings, construction plans and requirements of contracts, takes charge of materials purchasing and leasing to provide guarantee for the project; makes materials and equipments plans and implements them after they are approved by the manager in charge; sorts out and keeps all materials, equipment data, certificates and so on; sets up management account and does well the statistic work of material consumption and inventory; formulates the management standards of materials and equipments and implementation approaches, takes full charge of the quality and management of the materials and equipments of the project; controls project cost, enacts material dispensing limits, calculates mechanical cost and makes cost settlement.

8.2.2 Construction On-site Management

For high speed, safe and high quality finishing of the construction of the project, in addition to reasonable construction plan, technical measures and necessary construction equipments and people, greatest attention shall be paid to on-site management. Project on-site management level reflects one unit's whole quality and determines the realization of every goal. The project intends to adopt responsibility and goal management, namely, the management goals of the project will be divided layer by layer and implemented level by level. Each of the managements of construction has his or her own responsibilities and goals. Each of the responsibilities and goals at different levels of the project will be taken care of. Feasible regulations and systems shall be formulated to clarify the management rights and limit of each level and connect with individual income. The combination of responsibilities, rights and income shall be conducted.

8.3 Construction Solution of the Project

8.3.1 Main Construction Procedure

1. The first stage project (terrain moulding and biodiversity improvement project) Construction preparation—drainage—on-site clean up—setting out—concentrated sludge aeration—filling island side with the near earthwork of field ridges (including sample experiments by layer)—filling island center—the top covered with soil and gentle slop construction—other construction—completion and acceptance.

2. The second stage project (micro-topography wetland moulding and quick infiltration system project)
   Construction preparation—drainage—on-site cleanup—setting out—filling wetland with the bottom earthwork of ponds (including sample experiments by layer)—the filling of quick infiltration system—the top of small islands covered with soil and filled with stone and cinder and the construction of ecological bag slop and gentle slope—other construction—completion and acceptance.
3. The third stage project (restoration system project of forests, shrub wetland and shore belt wetland)
   Construction preparation—surveying and setting out—digging pits—planting trees—shore belt—planting emergent aquatic plants—other construction—completion and acceptance.

8.3.2 Construction Preparation and Temporary Facilities

Organization and management institutions shall be quickly established and improved after entering construction site; people will be quickly in place. Relationship of parties shall be coordinated; regulations and systems shall be perfected. Preparative work, such as temporary facilities construction, technical disclosure, measurement, positioning, earthwork balance calculation, etc.

8.3.2.1 Temporary Facilities of Production and Life

All temporary facilities are divided into two categories, life temporary facilities and productive temporary facilities. The former includes housing, offices, meet rooms, etc., which will be near leased houses of residents in the plan; the later includes material warehouses and equipment warehouses for which the near houses of residents or units will be rented as well as simple on-duty sheds that will be set up on the site. The former will be reduced in principle as far as possible by simplifying institutions and decreasing levels. All temporary facilities shall be constructed or rented in accordance with project progress and actual need.

8.3.2.2 On-site Leveling and Clean Up

All working places and land for life shall be leveled to guarantee normal construction. Low lying land shall be filled, heightened, compacted in association with earthwork digging; transverse ditches shall be established within the working places.

8.3.2.3 Electricity for Construction

Because of small electric load, the electricity for construction will be from near electricity grid. Besides, a 50 kW generating unit will be bought for drainage.

8.3.2.4 Water for Construction

The water for construction will adopt river water that has been tested and qualified. The water for life will adopt local running water. Transient drainage ditches will be digged on the construction site.

8.3.2.5 Temporary Roads

They will be main the roads in the construction site. When necessary, they will be constructed for the machinery in the construction site in accordance with the running lines of vehicles, construction setting out, construction organization.

8.3.2.6 Technical Preparation

Technicians shall be familiar with construction drawings and design intention and tell technologies to workers.

8.3.3 On-site Clean Up

Surveying and setting out shall be quickly carried out after entering the construction site; the inspection of all rivers, trees, tree roots, weeds, shrubs and floating items within construction area shall be conducted. The above matters within construction scope shall be cleaned up and put at appointed places.

8.3.4 Earthwork Digging and Refilling

8.3.4.1 Earthwork Digging

1. Before digging, the earthwork balance calculation and earthwork allocation plan shall be made in the principle of guaranteeing nearest distance. In construction, the earthwork shall be controlled with square control network in accordance with the requirements of the design drawings while attention shall be paid to the control of rolling quality and digging elevation.
2. After entering the construction site, water shall be drained. Bottom sludge depth shall be measured after drying in sunshine for 2–3 days. It is banned for machinery to directly enter into the sludge to avoid unexpected accidents.

3. Earthwork shall be digged from near places in line with construction setting out. The location of constructed islands can not be moved. Earthwork workload shall be reduced to save project cost.
4. In the course of digging earthwork, emphasis shall be laid on observing whether the water of surrounding fish ponds infiltrates into construction area. If any, it shall be handled.

8.3.4.2 Earthwork Refilling

1. Earthwork shall be pushed by bulldozers to near wetland, re-digged, filled by excavators and rolled layer by layer.
2. Before filling, all sundries must be removed, earth lumps whose diameter are more than 10 cm must be broken, the earth that include relatively greater water content shall continue to dry in sunshine to guarantee filling quality.
3. Before filling, the contractor shall choose the sample of the earth that will be used to conduct standard compacting experiment in accordance with SD28—84, and the water with best water content shall be used for filling.
4. The depth and compacting times of each layer of soil shall be dependent on soil property, water content, compacting coefficient and mechanical property. Rolling shall be overlapped; the joints between the uppers layer and lowers shall be staggered. The next layer can not be laid until one layer is inspected.
5. For filled soil, allowable settlement shall be reserved pursuant to design requirements; if no design requirement, 3 % of filled soil height shall be reserved.
6. The quality control contents of soil filling: Coordinates, height, levelness in level area; the central line location of digging and filling, section size and elevation; grade of side slope; filled soil compacting and dry density; hidden project record.
7. The allowable deviation the digging and filling of earthwork project: The bottom elevation of the foundation pit is 0–5 cm; the grade of side slope is in accordance with design requirement. The top elevation of the foundation pit shall be more than 1–5 cm (at the time of completion and acceptance). The dry density of the filled soil after compacting shall reach over 90 % in accordance with design requirements, the difference between the lowest value of the other 10 % and design value shall not more than 0.5 kN/m^2, and must not be concentrated.

8.3.5 Quick Percolation System Filling

The top height of small islands shall be reduced to 2.4 m, on which a layer of geotextile shall be laid first, then stone of 30 cm and cinder of 30 cm shall be laid, finally soil shall be filled to 3.5 m after laying level in accordance with design requirements.

8.3.6 Plant Planting

Plants shall be planted strictly pursuant to design to form special vegetation belts. Pterocarya stenoptera, metasequoia, taxodium ascendens, excoecaria sebifera, etc., shall be mainly planted on forest wetland belt; rose, forsythia viridissima, winter jasmine, etc., shall be mainly planted on shrub wetland belts; reed, cattail, wild zizania aquatica, etc., shall be planted on emergent aquatic vegetation.

8.4 Construction Progress and Guarantee System of the Project

8.4.1 Construction Progress and the Goals of Construction Period

Construction plan shall be made after entering construction site so as to guarantee construction period. The people, capital and materials of the construction plan shall be in place in time with good quality and sufficient amount in accordance with the requirements of the construction plan; and the project progress will not be affected due to this. The planned construction period of the first stage project is 90 days, the second stage project is 45 days, and the third stage project is 150 days.

8.4.2 Quality Guarantee System and Guarantee Measures of the Project

The quality guarantee system of the project is seen as Fig. 8.2.

Quality guarantee measure of the project.

1. Quality management network headed by project manager and taken charge of by chief engineer shall be established, quality guarantee plan and quality goals and policies shall be clarified, and ISO 9000 shall be implemented. All-round quality management shall be carried out, and the quality management activities involved in all staff and whole process shall be proceeded. Quality shall be continuously improved through PDCA cycle, construction shall be conducted strictly in accordance with specifications and design requirements; layer by layer technical disclosure and level by level quality control shall be carried out in accordance with the technical plan of construction.
2. The work of quality department and technical department shall be taken charge of by chief engineer. The on-site inspection system of the project shall strict; the right of "quality vetoed by one vote" shall be given to quality inspectors. Dynamic management shall be exercised with attention paid to

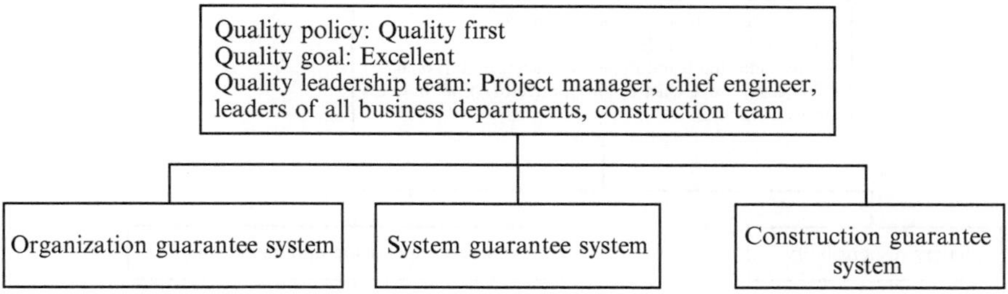

Fig. 8.2 Quality guarantee system

process management. Prevention shall be emphasized, and quality management points shall be set up at key locations, key working procedures and special working procedures. Task guidance pamphlets shall be complied to made clear of standards and responsibilities; next procedure shall not begin until one procedure is inspected and qualified by quality inspectors and supervisors.

3. Joint review system of drawings, level by level technical disclosure system, material inspection system and quality analysis and report system shall be established and strictly followed to ensure technically guaranteeing construction.
4. The work of self-inspection, mutual inspection and handing in inspection and special inspection shall be done well.
5. Central laboratory shall be set up in project manager department, equipped with professional testers. Sampling inspection shall be performed in accordance with the requirements of specifications; mixing proportion that is qualified and meet the requirements of construction shall be designed in line with indexes. In addition, the laboratory shall bear the inspection items prescribed in specifications to provide necessary data for guiding construction.
6. Measurement and setting out shall be conducted using total station, transit and level and strictly in accordance with central axises, traverse points, stake marks and bench marks. Whole line joint measurement of whole line measurement control points shall be periodically taken, and joint measurement results shall be analyzed and reported to supervision engineers for approval to facilitate modification.
7. Importance shall be given to quality guarantee and data management; the construction quality guarantee data shall be carefully filled in, sorted out and filed.
8. Quality education shall be strengthened to improve the quality awareness of all employees by usually proceeding labor competition activities of "quality in my heart". Strict and perfect quality award and punishment system shall be formulated to link project quality with economic benefit to guarantee the project quality with economic means.
9. For key and difficult projects, the construction managements, technicians and backbones of teams who have construction experiences in similar projects shall be input.

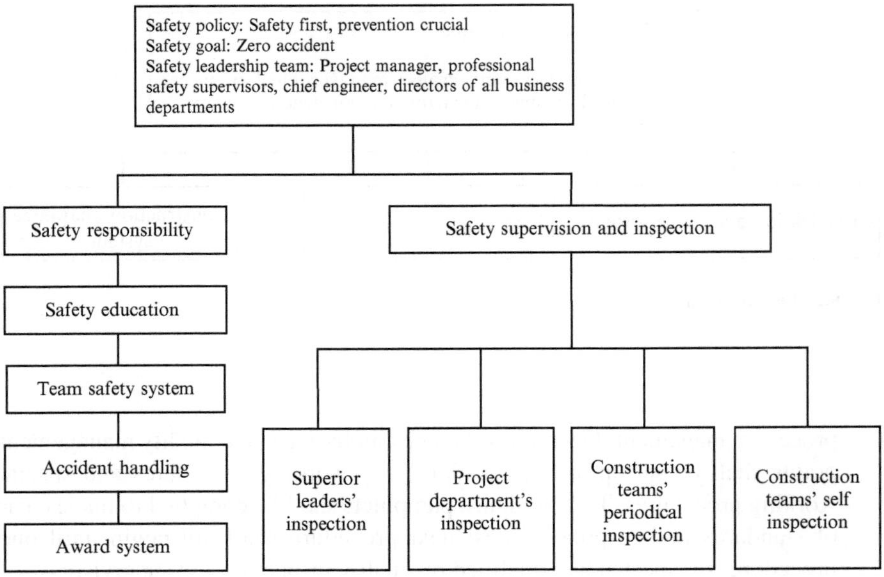

Fig. 8.3 Safety guarantee system

10. The quality of seedling plants shall be controlled strictly pursuant to design requirements; great efforts shall be made to conduct quality inspection and control in planting seedling plants, especially, in planting trees. Rain forecast and drainage shall be intensified to guarantee the quality and construction period of the project.
11. Technical guarantee measures of construction: The guidance of all the kinds of work of the project is to cerate a high quality project. The following system shall be exercised in order to ensure that the whole project is technically guaranteed: Technical review and hidden project acceptance system, technical quality disclosure system, three-level acceptance and project quality appraisal by department and by item system.

8.4.3 Safety Guarantee System and Safety Guarantee Measures

The safety guarantee system is seen as Fig. 8.3.

The safety guarantee measures of the project:

1. Safety management network shall be set up; three-level safety protection responsibility system shall be carried out. The management leadership team headed by project manager and safety protection section with professional safety supervisors responsible for safety management of the project shall be established under the project manager department. Part time safety supervisors

shall be employed in each team, responsible for the safety inspection and supervision of construction on-site sections. Efforts shall be made to clarity responsibilities and ensure safety.

2. The policy of "safety first, prevention crucial" shall be stuck to, and the right of "safety vetoed with one vote" shall be given to safety supervisors. Safety education shall usually be strengthened to make safety work systematized and regular. Safety education and safety technical disclosure shall be conducted in an all-round way before construction beginning to improve employees' safety awareness.
3. Inspection, summary and construction record shall be carried out during construction. Project manager shall inspect periodically, special safety supervisors shall supervise and inspect at each shift, and the safety supervisors of each team shall inspect on-site safety condition 10 min before work. Safety inspection diary shall be kept; potential safety hazard shall be carefully inspected.
4. New constructors shall be educated and trained in safety production. They can not enter operation posts before they are examined and qualified.
5. Safety helmets shall be worn before entering construction site; protective rales and safety net shall be set up in pouring.
6. Protective devices shall be equipped on electrical facilities of construction; electricians shall often check electrical equipments to ensure safety without accident.
7. Fire protection education shall be intensified; "three treasures and one equipment" shall be equipped on construction site to prevent accidents from happening.
8. All kinds of mechanical equipments shall be uniformly managed and dispatched. The operation against rules shall be strictly banned to guarantee traffic safety.
9. Construction at night must be conducted under sufficient lighting, directed by specially-arranged people to ensure safety.
10. Safety activities day once a week shall be set up. Safety production activities shall be inspected before and after shift and at meetings. Safety education and publicity shall often be conducted to employees.

8.4.4 Technical Measures of the Construction in Winter and Monsoon

Because the construction of the project is concentratedly conducted in winter and monsoon, how to resolve the problems of anti-freezing, rainfall and drainage is the key in term of project progress and quality.

Filled earthwork collapse due rainfall immersing shall be prevented by prevention before rain and action after rain.

Filling shall be loosened and dry in sunshine if it has great water content after rain.

Constructors shall be equipped with winter clothes and rain coats to prevent flu and other epidemic diseases from happening.

8.4.5 On-site Civilized Construction Measures

Civilized construction is especially important because the construction site is located in suburb.

1. The civilized construction management and organization institution headed by the project manager and consisting of the people of each department and section and production team shall be established to create an environment of civilized construction.
2. Clean and civilized construction site shall be realized by strengthening education and publicity, improving employees' awareness of civilized construction and hanging "four signs and one figure".
3. Regulations and systems shall be enacted to strengthen inspection supervision.
4. Clean construction site shall be kept by reasonably arranging construction site, rationally setting facilities, tidily piling materials and equipments, banning private and random connection. The sanitation along the roads and construction area shall be guaranteed when earthwork is transported. Attention shall be paid to civilization, courtesy, integrity and fighting against ill actions.

8.4.6 Construction Management and Others

The project is characterized with short construction period, great construction difficulty and high requirements of quality, so it must be carefully organize, scientifically manage and set up and improve management networks of all kinds. Modern operation and management knowledge shall be used in every field and link. Importance shall be given to the management work of production, quality, materials, equipments, etc. The safety management shall be put in the first place. Management shall generate benefit and promote project progress. The project shall be done with first class quality, first class speed and in an all-round way.

8.5 Supervision of the Project

8.5.1 Supervision Goals and Guidance

8.5.1.1 Supervision Goals

With the tenet of strengthening construction management, improving construction quality, fully exerting investment return, the supervision goals of the model project of the model area of Dalian Lake wetland restoration shall contain:

1. Construction period goal: The supervision center will follow the proprietors' overall construction period plan and arrangement to organize supervision department enter construction site in time and reasonably arrange and dispatch supervisors to guarantee that supervision will goes smoothly. People shall be arranged in the principle of stability, excellence and suitableness. The supervision department will provide services from the beginning to the end. It guarantee fulfilling the construction period goal based on the construction goal specified in the construction contracting contract of the project and controlling the construction period of each construction section strictly in accordance with the requirements of the construction period of the construction contract.
2. Investment goal: The contract amount agreed by the project owner and contractor will serve as the basic goal of supervision and investment control. The actual payment of all kinds of costs shall be continuously monitored in the course of construction by carefully signed certificates, correctly handling change and claim.
3. Quality goal: The project shall be controlled to be qualified 100 %, so shall the projects of all departments. The project shall be uniformly divided, reported to the proprietor and implemented after approval.
4. Organizing and coordination: Contradictions and problems will arise inevitably due to the project owner and contractor's different economic interests and different understandings of problems. The supervisors shall coordinate and resolve them in time and fairly and protect the legal rights and interests of each party of the contract pursuant to the contract. At the same time, they will carry out external coordination together with the project owner to provide necessary working environment and external condition.

8.5.1.2 Supervision Guidance

Efforts will be made to achieve the overall goal of the construction contract of the project in an all-round way through the organization, coordination, supervision, control and services. Every goal of the project will be completely fulfilled by independent and just standpoint, strictly abiding by the supervision principle of "just, lawful, scientific and credit", through "three controls", "two managements" and "one coordination". For this, the following will be done:

1. Adherence to all-round economic benefit. Based on related laws and regulations and national macroeconomic benefit and social benefit, the supervisors will safeguard the microeconomic benefit of the investment project of the project owner.
2. Sticking to high standard services. The rights and duties specified in the supervision contract shall be strictly exercised, consistence of rights and responsibilities shall be kept to, and every effort shall be made to strict, careful and high efficient supervision. Strict requirements will be imposed on the contractor. Integration of supervision and assistance shall be carried out. The measures of economy, contract,

technologies, organization, etc., will be flexibly employed to guarantee achieving project quality, progress and investment goal and create high quality model project and outstanding supervision achievements.

3. Keeping justice. The legitimate rights and interests of both the project owner and contractor shall be safeguarded equally pursuant to the contract in the whole course of project management. Based on sciences and objective facts, a dispute will be justly coordinated in the independent identity of the third party when it arises between the two sides.
4. Scientific management mechanism. All-round scientific project management system or supervision and control procedure shall be established; emphasis shall be given to the control in advance in supervision with the concept of "predicting before the event, taking measures in the event, summarizing after the event" to ensure supervision effect. Scientific construction and strict supervision shall be adhered to.

8.5.1.3 Supervision Basis

National and local current related laws and regulations:

1. The approval documents and design documents of the project and the tender inviting documents of the project owner;
2. *Project Construction and Supervision Contract* signed by the project owner and supervisor;
3. The construction contracting contract signed by the project owner and contractor and other documents;
4. Construction drawings and their statements;
5. Contract projects list and statements;
6. The standard drawings, technical specifications, project quality inspection and appraisal standards, experimental regulations, etc., that are specified in contract for use.
7. Other contracts that form part of the contract.

8.5.1.4 Supervision Scope and Contents

The supervision scope is the supervision in the construction period of the model project of the model area of Dalian Lake wetland restoration. The supervision contents are mainly construction period quality control, progress control, project cost control, contract management, information management, etc.

8.5.2 Supervision Institution

Targeted at the characteristics of the project, Zhenjiang Huayuan Construction Supervision Center intends to establish Zhenjiang Huayuan Construction Supervision Center Supervision Department for Dalian Lake Wetland Restoration Mode Project.

To guarantee the smooth implementation of the supervision work and its need for supervisors and carry out the principle of "excellence and high efficience", the supervision institution will mainly consist of young and middle-aged people to meet the requirements of professional supervision.

8.5.3 *Task and Method of Investment Control*

8.5.3.1 The Tasks of Investment Control

1. To make contract payment, check and sign project payment certificate in accordance with approved projection construction progress, plan, divisions of goals and plans;
2. To adjust contract price pursuant to the provisions of the contracting contract of the project and the proprietors' authority;
3. To assist the proprietor in project completion settlement.

8.5.3.2 The Method (Measure) of Investment Control

The measures of organization, economy, technologies, contracts and so on, shall be taken to control investment.

Measure of contract:

1. Construction units shall be requested to take construction record, especially, the change in the course of construction, to provide basis for correctly handling possible claim.
2. To participate contract modification and supplement, to pay attention to its affection on investment control.

8.5.3.3 Specific Methods of Investment Control

Project Metering

In accordance with the provisions of the contracting contract of the project, the basis, principles, methods, procedures, review and correction of project metering shall be determined. A whole set for working flow of project metering shall be enacted.

Project Payment

In accordance with the provisions of the contracting contract of the project and technical terms, project payment management procedure shall be made, the declaration conditions, declaration contents and requirements of project payment shall be determined, and the inspection of project payment declaration, signing of project payment certificates and its modification shall be conducted.

8.5.4 *Tasks and Methods of Progress Control*

In addition to the progress control in accordance with the overall construction period goal specified when the proprietor signed construction and supervision contracts, the construction unit shall be required to determine the construction periods of the sections of the project after entering construction site to facilitate the progress control of the project.

8.5.4.1 The Tasks of Progress Control

The tasks of progress control are mainly:

1. Review of the progress plan provided by the contractor;
2. Control of progress course;
3. Handling of the application of project delay claim of the contractor in line with the provisions of the contracting contract;
4. Provision of the optimizing proposals and analysis reports about construction progress and construction period to the proprietor;
5. Report of project progress information to the proprietor in accordance with the provisions of the construction and supervision contracts of the project.

8.5.4.2 Methods of Progress Control

1. To assist the contractor to implement progress plan

The supervision department shall pay attention to knowing the questions in the course of the implementation of the progress plan at any time and help the contractors to resolve it, especially the coordinative problem of internal and external relationship.

2. To hold special coordinative meetings of project progress

The supervision department shall regularly or irregularly hold the special coordinative meetings of project progress to settle the coordinative questions in the course of construction. In case of emergencies, the supervision department shall issue urgent coordinative orders to urge related units to adopt emergent measures to safeguard normal construction order.

3. To issue project progress payment certificates

The supervision department shall verify the projects reported by the contractor (the projects of the departments of the project can not be reported until they are finished), and then issue project progress payment certificates after check and acceptance.

4. To summit progress reports to the proprietor

The supervision department shall sort out progress data every month, make project record and summit progress reports to the proprietor periodically.

8.5.5 The Tasks and Methods of Quality Control

8.5.5.1 Project Quality Supervision

The main tasks of quality control shall be as follow.

Review of starting construction. After the proprietor signs construction contract with the contractor, the supervision department shall make preparation before the contractor enters the construction site and strictly review whether the contractor has met the conditions of starting construction after he enters the construction site. The supervision department strictly inspects the conditions of starting construction (people, equipments, etc.) before the project of each unit starts construction.

8.5.5.2 The Methods of Quality Control

1. Reasonable division of the project. The project division form of quality inspection and appraisal shall be made in accordance with the project and associated with quality appraisal standards. The quality appraisal form formats of the project shall be made to standardize and formalize quality control and inspection.
2. Enacting quality control standards. Quality control standards shall be enacted by reference to quality appraisal standards, construction workmanship and the technical terms of the contract, and continuously improved by association with construction situation so as to meet the requirements of quality control of the project.
3. Strict review of technical reports. The method is mainly to review the formal report of starting construction provided by the constructor, the technical measures of the key parts of the project and the technical identification report about new materials and new technologies, etc.
4. Monitoring and inspection. Supervisors shall be arranged to monitor and inspect construction pursuant to the requirements of the quality control of the project. The action against operation specifications shall be stopped in time to avoid quality accidents. The methods of quality monitoring and inspection are sensory inspection, survey inspection, material experiments and quality sampling inspection, etc.
5. Strict quality acceptance. The quality certification right and veto right endowed by the contractor shall be correctly exercised. The next procedure shall not be allowed to be carried out until its front procedure is accepted and qualified. The unqualified parts of the project shall by no means be metered and paid, and the contractor shall be required to rework, repair until they are qualified.

8.5.6 The Supervision Method of Defect Responsibility Period

The supervision department shall, pursuant to the supervision time, scope and contents specified in the supervision contract, arrange supervisors to inspect and record the quality defects of the project proposed by the proprietor and accept the project quality repaired by the contractor. The supervisors shall investigate and analyze the reason of the quality defects and determine who shall take the responsibilities. For the quality defects not caused by the contractor, the supervisors shall verify the repair cost and sign project payment certificates and report them to the proprietor.

Part III
Effect Monitoring and Appraisal

Chapter 9
Wetland Water Quality Monitor and Appraisal

9.1 The Water Quality Monitoring Layout and Appraisal Methods of the Project

9.1.1 Monitoring Layout and Time

The sampling and monitoring of the water quality and planktons of the wetland after the first stage model project of Dalian Lake were carried out between April 2010 and February 2011. Nine sampling places (No. 1–9) were set up in the whole project area, five (No. 10–14) were established at outside rivers before sludge was removed. Later, the sampling place No. 15 was added to Lanlu Port in accordance with the real condition. Sampling places are seen as Fig. 9.1.

9.1.2 Water Quality Monitoring and Appraisal Methods

9.1.2.1 Water Quality

Water is collected with a glass water collector. The two semi-circular upper covers of the instrument can easily be opened and closed; there is a hole and floating board at its bottom which can guarantee that water can freely come in and go out when it sinks into water; there is a thermometer in its inner wall and a water outlet in its exterior wall. When the water collector is put into water, the movable lower door at the bottom opens by the virtue of upward impact, and then water enters it. As it continues to sink, water continues to enter it. It is lifted after water reaches sampling depth, and then the upper and lower doors shut at the same time. Thus, the purpose of sampling is achieved while water temperature is also obtained. Then, the temperature is recorded. Finally, the sample water is poured into a water collecting bottle and kept airtight in dark.

S. An and L. Wang, *Wetland Restoration: Shanghai Dalian Lake Project*,
Springer Environmental Science and Engineering, DOI 10.1007/978-3-642-54230-5_9,

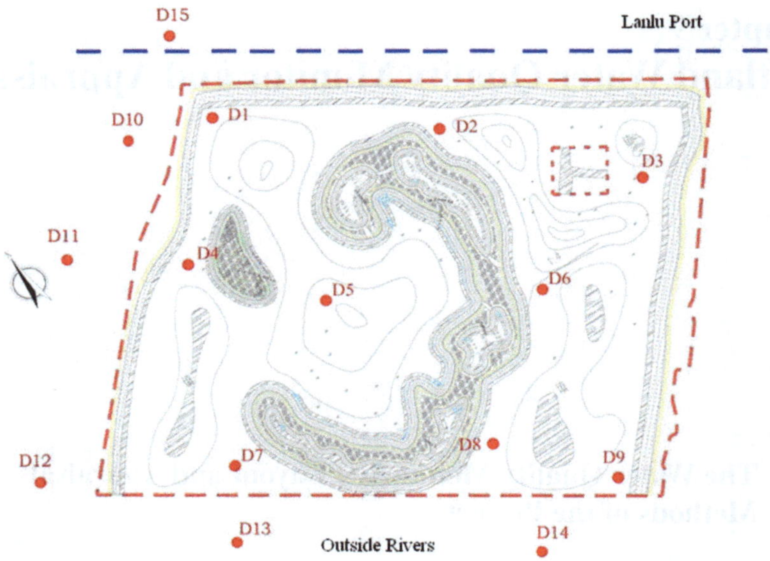

Fig. 9.1 The sampling places of the water quality monitoring of Dalian Lake

Three bottles of sample water is taken at each of the sampling places in Fig. 9.1 in accordance with the above method respectively for determining the physical property (SS%, Temp, DO, etc.) and chemical property (pH, COD, TN, NH_4-N, NO_3-N, TP) of the water; and the depth and transparency of the water are determined while the sample water is taken.

Besides, three bottles of sample water is taken with the same method for determining the content of chlorophyll *a* in the water.

9.1.2.2 Planktons

1. The qualitative samples of planktons: Planktons are collected with a plankton net. The plankton net ties to a bamboo pole and check whether the switch on the top of the net is closed. Switch on the switch to clean the net, and then switch it off. Put the net into water with the bamboo pole in hand; 5 min later, slowly lift the net. All planktons concentrate on the top of the net after water leaks. At that time, put a small bottle used for keeping plankton samples under the switch, switch on the switch to let the plankton samples flow into the small bottle. Phytoplanktons are collected with plankton net No. 25 and zooplanktons with plankton net No.13. The specimen needs to be fixed by adding stationary liquid into the specimen bottle. Phytoplanktons are fixed with 15 over 1,000 Lugol's solution, and zooplanktons with 2–4 % formaldehyde. Then tags are stuck to the bottles and record is made.

2. The quantitative samples of planktons: The samples are collected with a water collector at the water depth of 0.5 m. Then they are immediately fixed. Paste tags and take record. 1 L sample water is collected for phytoplankton, more is collected due to need for fixing volume; and then 15 mL Lugol's solution is immediately added to it. 10 L sample water is collected for zooplanktons, filtrated through a plankton net and put into a jar. Finally, add 2–4 % formaldehyde, paste the tag and take record.

9.2 The Monitoring Results of the Water Quality of the Project

9.2.1 The Physical Property of the Water

The physical property of the water is determined with multi-parameter water quality analyzer.

9.2.1.1 Transparency

The transparency is determined with plug's plate method.

It is a method to determine transparency on spot, namely a disk with white and black color of diameter 20 cm is put into water, the depth at which the disk is just not seen expresses transparency.

The transparency means the clarification extent of sample water. Turbid water affects the sense of water and is one of the symbols that water is probably polluted. The water with low transparency obviously hinders light transmission and affects the subsistence of aquatic organism. It can be seen from data analysis (Fig. 9.2),

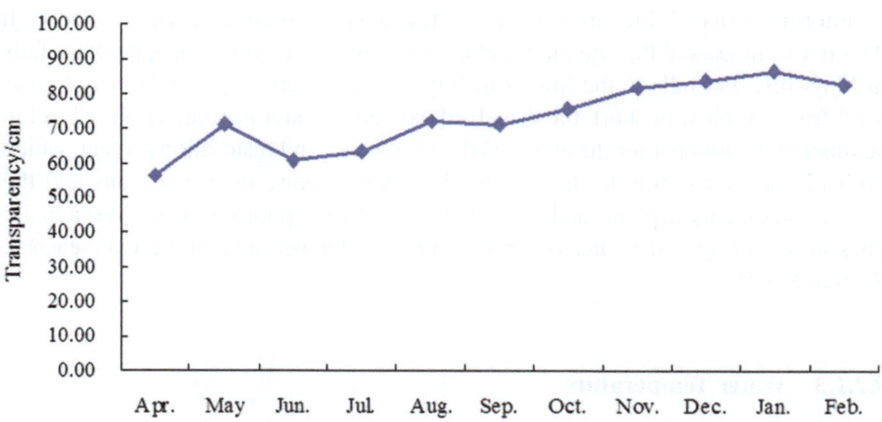

Fig. 9.2 The change trend of the water transparency of Dalian Lake

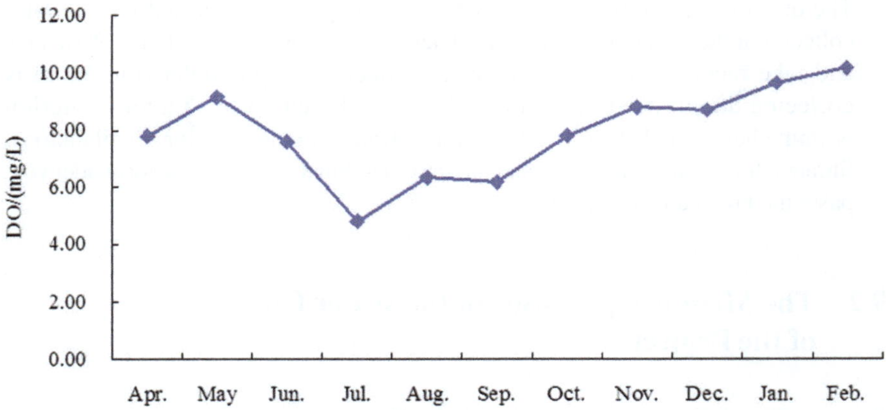

Fig. 9.3 The change trend of the dissolved oxygen in the water of Dalian Lake

the water transparency of Dalian Lake took on the trend of obviously going up after the end of the restoration project from April to May. This also indicates that the water quality has been purified to some extent after wetland restoration. But the transparency began to slightly fall from May, reached the lowest value 61.4 cm by June and July, but clearly higher than the value measured in the first month after wetland restoration. Although there is not blue-green algae breakout in summer within the scope to Dalian Lake project, June and July must be the period in which all kinds of algaes multiply in great quantity, which causes the decline of water transparency.

9.2.1.2 Dissolved Oxygen

Clearly, the dissolved oxygen in water undergoes a relatively uniform change in monitoring period. It increases in April–May, which is mainly closely related with the photosynthesis of the aquatic vegetation within the project area; apparently falls in May–July and falls to the lowest in July; gradually goes up to relatively higher level from October. In short, the dissolved oxygen in water is relatively low level in summer and autumn after the end of Dalian Lake wetland restoration project, which probably has something to do with the high temperature in two seasons and the respiration of aquatic plants and algae that exceeds their photosynthesis; it gradually goes up in spring and winter because of the low temperature in the two seasons. Seen in Fig. 9.3.

9.2.1.3 Water Temperature

The even value change trend of the water temperature at the sample places 1–9 is seen in Fig. 9.4. During the period from April 2010 to February 2011, water temperature

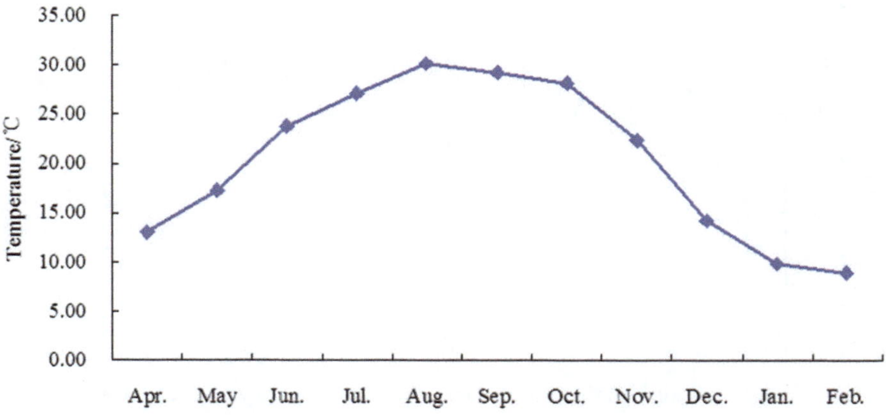

Fig. 9.4 The change trend of the water temperature of Dalian Lake

gradually began to go up from April 2010, reached the highest value 30.22 °C in August, and then gradually went down, reached the lowest value 8.91 °C in February 2011.

9.2.2 *The Chemical Property of Water*

9.2.2.1 pH Value

The pH value in natural water depends on free carbon dioxide content and carbonic acid balance.

In accordance with the formula CO_2 (dissolved in winter) + $H_2O \rightleftharpoons HCO_3^- + H^+ \rightleftharpoons CO_3^{2-} + 2H^+$, if other conditions remain unchanged, the more CO_2 (dissolved in winter), the lower pH value. Aquatic organisms have relatively great affection on pH value; their aspiration then can give out great amount of CO_2 makes pH value fall while aquatic plants (including algae) whose photosynthesis can absorb CO_2 makes pH value go up. At night, photosynthesis stops, organisms and algae whose aspiration produces great amount of CO_2 makes pH value go down. The pH value in water changes with the change of season and biological vertical distribution because of the role of algae and fungus (Fig. 9.5).

Obviously, it can be seen from the monitoring results that the pH value of the water in Dalian Lake go up in April–June, with maximum value 8.45, because the temperature in spring is relatively low at the beginning, the CO_2 concentration of water is relatively high, later, as aquatic plants grow, photosynthesis exceeds aspiration, which leads to the fall of CO_2 dissolving in water. In July–October, as the dissolved oxygen in water decreases, the aspiration of the aquatic plants exceeds their photosynthesis, the CO_2 concentration of water increases; so pH value goes down, with lowest value 7.91. After November, due to the gradual fall of temperature, the death of aquatic plants and other reasons, pH value changes accordingly.

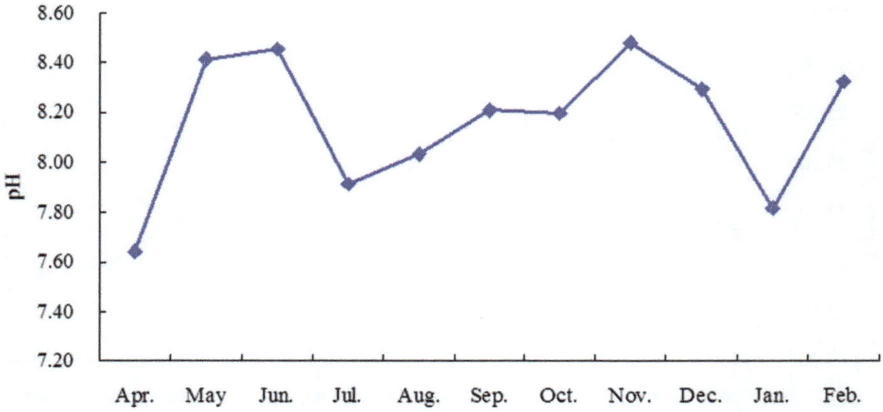

Fig. 9.5 The change trend of the pH value of the water in Dalian Lake

9.2.2.2 Permanganate Index COD

After collected, the samples of determining chemical oxygen demand (COD) are sent to laboratory and must be tested within 3 days.

Sample Water Analysis

100 mL sample water is put into a triangular bottle of 250 mL (if the organic content in the sample water is relatively high, moderate amount of the sample water is taken and diluted to 100 mL with pure water). Add 5 mL sulfuric acid ($V_{H_2SO_4} : V_{H_2O} = 1:3$) and 10.0 mL per manganate [$c(1/5KMnO_4) = 0.01$ mol/L] then shake up. Put the triangular bottle on an electric stove to heat until the water boils. And then immediately put the bottle into boiling liquid (higher than the sample liquid in the bottle) to be heated for 30 min. Later, take out the bottle, add 10.0 mL sodium oxalate standard solution to it and shake up. After the fuchsia of permanganate disappears (at that time, the sample liquid should not be lower than 70 °C, otherwise, it needs to be heated), permanganate standard is dripped into the sample liquid until the slightly red color does not fade.

Result Calculation

Sample water COD is calculated in accordance with the following formula (expressed with mg/L oxygen consumption):

$$\rho_{O_2} \text{ or } \rho_{COD} = \frac{C \times (V_1 - V_2) \times 7.9997 \times 1{,}000}{V}$$

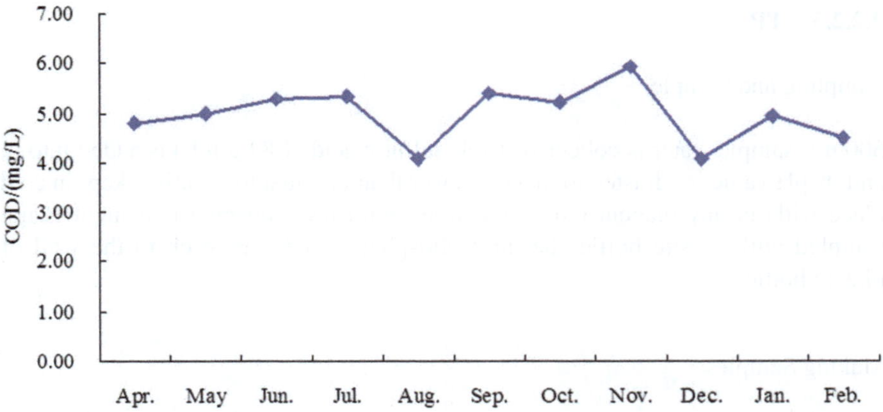

Fig. 9.6 The change trend of the permanganate indexes of the water in Dalian Lake

Table 9.1 The standard limits of the basic items of surface water environment quality standard (Unit: mg/L)

Item	Class I	Class II	Class III	Class IV	Class V
Permanganate index	≤2	4	6	10	15
Ammonia nitrogen (NH_3-N)	≤0.15	0.5	1.0	1.5	2.0
Total phosphor (on P basis)	≤0.02	0.1	0.2	0.3	0.4
Total phosphor (lake and reservoir, on N basis)	≤0.2	0.5	1.0	1.5	2.0

Where, ρ_{O2} or ρ_{COD}: mass concentration of sample water COD, mg/L-O_2; C: concentration of permanganate standard solution, mol/L; V_1: volume of permanganate standard solution added in the course of testing, namely the volume of 10 mL plus the drip, mL; V_2: volume of permanganate standard solution equivalent to 10.0 mL sodium oxalate standard solution.

Result Analysis

The change trend of the permanganate index event values of the sample water collected at sample places (1–9) is seen in Fig. 9.6. The permanganate index is the important index of the monitoring of surface water, drinking water and domestic sewage and reflects the pollution extent of organic and inorganic oxidable materials. It can be seen from the figure that between April 2010–February 2011, the COD indexes of the water in Dalian Lake changed little, the overall even value fluctuated between 4.0 and 6.0 mg/L.

It can be seen from national surface water environment quality standard (Table 9.1) that in term of permanganate index, the water in Dalian Lake is in the state of Class III water almost whole year. This shows that the water in Dalian Lake wetland has improved to certain extent after wetland restoration project.

9.2.2.3 TP

Sampling and Sample

500 mL sample water is collected, 1 mL sulfuric acid (1.84 g/mL) is added into it, and its pH value is adjusted to make it lower than or equal to 1, or it is kept in cold place without any reagent (Note: the water with less phosphorus content is not sampled with plastic bottles because phosphate is easy to stick to the wall of plastic bottles).

Making Samples

25 mL sample water is carefully collected so as to get the sample water whose dissolved part and suspended part are typical and put into scale pipe with stopper. If its phosphorus content is high, sample water volume can be reduced.

Potassium Persulphate Dissolving

4 mL potassium persulphate (50 g/L) is added to the above sample water, the scale pipe with stopper is lidded, its glass stopper is tied up with a small piece of cloth and thread (or use other methods to fix), and then it is put in a big beaker which is put in a high pressure steam sterilizer to be heated and kept for 30 min after the pressure reaches 1.1 kg/cm^2, and temperature reaches 120 °C, taken out for cooling and diluted with water to scale line.

Color

Add 1 mL ascorbic acid solution (100 g/L) respectively to each portion of diluting solution and shake up; 30 sec later, add 2 mL molybdate solution and shake up.

Spectrophotometry

After laying aside for 15 min, absorbance is determined using the cuvette with optical length 30 mm, under wavelength 700 nm and by reference to water. Phosphorus content is obtained from the standard working curve after deducting the absorbance of blank test.

Result Analysis

The change trend of the total phosphor index even value of the sample water collected at sample places 1–9 is seen in Fig. 9.7. The total phosphor of the water is in decline trend between April and July due to the gradual restoration of the wetland

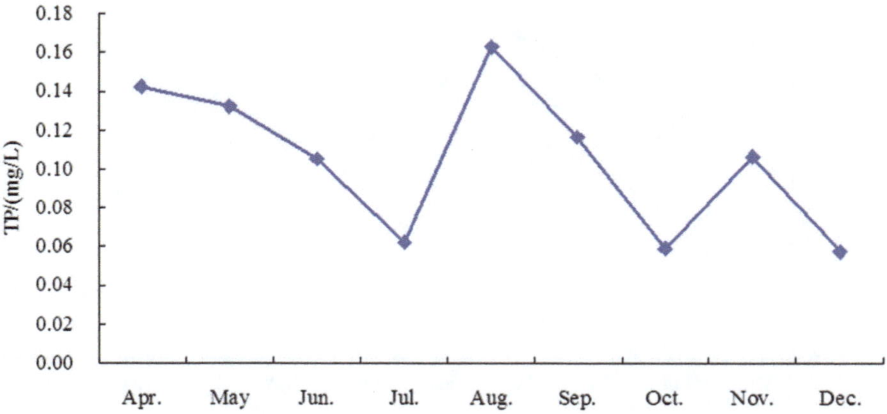

Fig. 9.7 The change trend of the total phosphor of the water in Dalian Lake

and being in high flow period. But most of the total phosphor standard limits are higher than the total phosphor standard limit 0.1 mg/L of Class II water and lower than the total phosphor standard limit 0.2 mg/L of Class III water (Table 9.1). So the water in Dalian Lake belongs to Class III water. This is the result from the restoration project. But the total phosphor suddenly go up to 0.163 mg/L in August and maintains relatively high level in September, which is the time when algae begins to die and decompose. The rich total phosphor that algae itself contains is released in a short period of time, which results in the increase of the total phosphor. On the other hand, Dalian Lake belongs to the shallow lake. Its bottom sludge is easy to be suspended because of the role of the wind wave in summer, leading to the release of the nutrients in particles. The total phosphor gradually reduces between October–December.

9.2.2.4 TN

The change trend of the total nitrogen index even value of the sample water collected at the sample places 1–9 is seen in Fig. 9.8. Obviously, the total nitrogen is in decline trend on the whole. It can be seen in accordance with national surface water environment quality standard (Table 9.1) that the total nitrogen indexes of the water in Dalian Lake are nearly in the standard of Class II water, more than 0.5 mg/L between July and September, reduce to the standard of Class III. The total nitrogen includes inorganic nitrogen and organic nitrogen. The former refers to NH_4^+-N, NO_3^- and NO^{2-}, etc. Algae main absorbs NH_4^+-N. The nitrogen in the water of Dalian Lake is different in different seasons and months. In spring, when the phytoplankton, large sized aquatic plants begin to multiply in great quantity and grow, which is in high flow period, the total nitrogen reduces. Later, it begins to go up and reaches the highest value in August. It goes down again after August until about January.

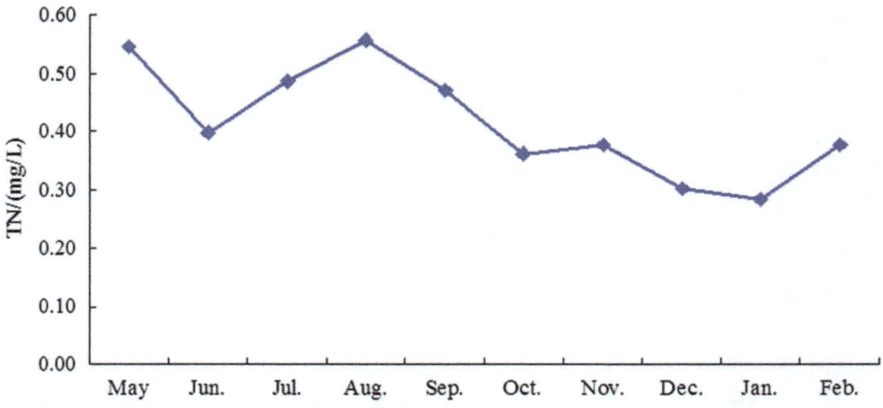

Fig. 9.8 The change trend of the total nitrogen of the water in Dalian Lake

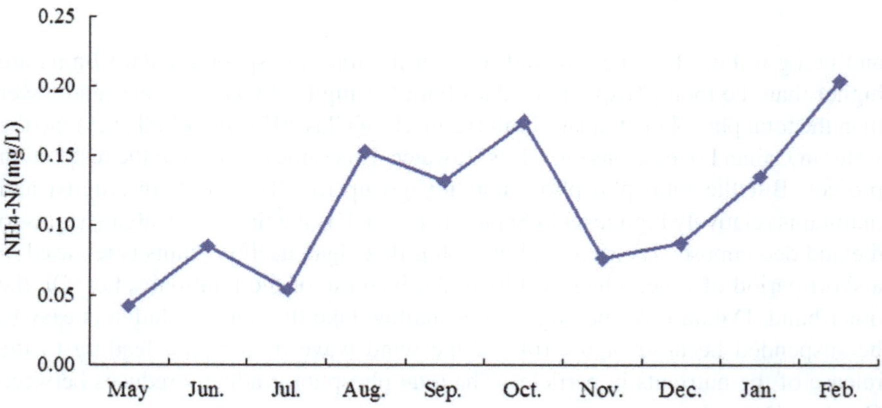

Fig. 9.9 The change trend of the ammonia nitrogen of the water in Dalian Lake

9.2.2.5 Ammonia Nitrogen

Algae grow mainly dependent on absorbing NH_4^+-N in total nitrogen. So the ammonia nitrogen values slightly increase in May and June, gradually decrease in July as algae grows and multiplies, gradually go up after July, but all are lower than the ammonia nitrogen standard limit 0.5 mg/L of Class II water. This is a testimony of the improvement of the water quality. Seen in Fig. 9.9.

9.2.2.6 Nitrate Nitrogen

As shown in Fig. 9.10, the nitrate nitrogen takes on the trend of slightly going up on the whole, with relatively great increase in August, which has something to do with the death and decomposition of algae. It reaches peak value in August and gradually goes down since then, but it is still the trend of going up on the whole.

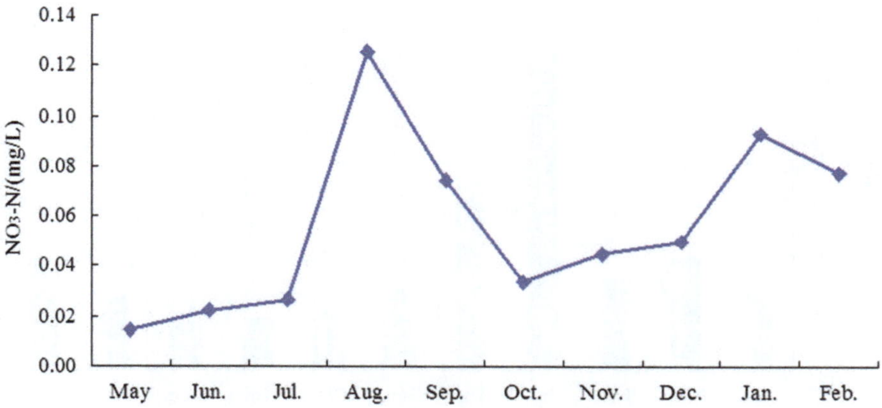

Fig. 9.10 The change trend of the nitrate nitrogen of the water in Dalian Lake

9.2.3 The Biological Nature of Water

9.2.3.1 Chlorophyll

The Determining Method of Chlorophyll

1. Take moderate amount of sample water, add small amount of magnesium carbonate powder to it, reduce pressure and filter through filter membrane to intercept the phytoplankton cells of the sample water.
2. Put the filter membrane into fridge for low temperature drying, then grind and extract 24–36 h the filter membrane with 90 % acetone.
3. Separate filtering liquid centrifugally to abstract upper clear liquid constant volume 10 mL colorimetric cylinder.
4. Determine absorbance value in 1 cm cuvette, by reference to 90 % acetone, at the wavelengths of 750, 663, 645, 630 nm of spectrophotometer and then calculate chlorophyll content in accordance with the following formula.

$$\text{Chlorophyll } a\ \left(mg/m^3\right)=\left[11.64\left(A_{663}-A_{750}\right)-2.16\left(A_{645}-A_{750}\right)+0.10\left(A_{630}-A_{750}\right)\right]\times V_1/\left(V\times C\right)$$

Where, C: cuvette optical length(cm); A: absorbance; V_1: volume after abstracting liquid constant volume (mL); V: volume of sample water (L).

Result Analysis

The change trend of the chlorophyll even values of the sample water taken at sample places 1–9 of Dalian Lake is seen in Fig. 9.11. The chlorophyll even values of the water are all lower than 2.0 μg/L between April–June, obviously go up in July with

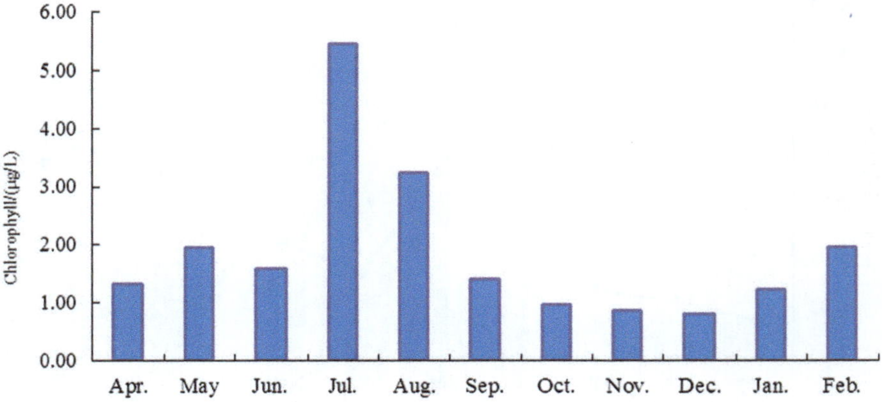

Fig. 9.11 The change trend of the chlorophyll content and nitrogen of the water in Dalian Lake

the maximum value 5.44 μg/L. The chlorophyll content in the water of the lake maintain a relatively high level in July and August, which is related to the growing and multiplying in great amount of blue-green algae. It can be seen from the Fig. 9.11 that the chlorophyll content gradually increases in January.

9.2.3.2 Phytoplankton

The Type Identification and Quantity Determining Method of Phytoplankton

The types of phytoplankton are identified by reference to *Freshwater Microorganism Atlas* and *Freshwater Plankton Atlas*.

The quantity of phytoplankton is calculated with 0.1 mL account chamber with area 20 mm × 20 mm and volume 0.1 mL which is internally divided into ten horizontal lines and ten vertical lines, a total of 100 checks. Phytoplankton is calculated with 20 of them.

The density calculation formula of phytoplankton is:

$$N = n \times (A \times V) / (AC \times V_a)$$

Where, N: number of phytoplankton in per liter of original sample water (piece/L); A: volume of accounting chamber (mm^2); AC: Area (mm^2); V: volume of 1 L original sample water after sediment and compression (mL); V_a: volume of account chamber (mL); n: number of accounted phytoplankton.

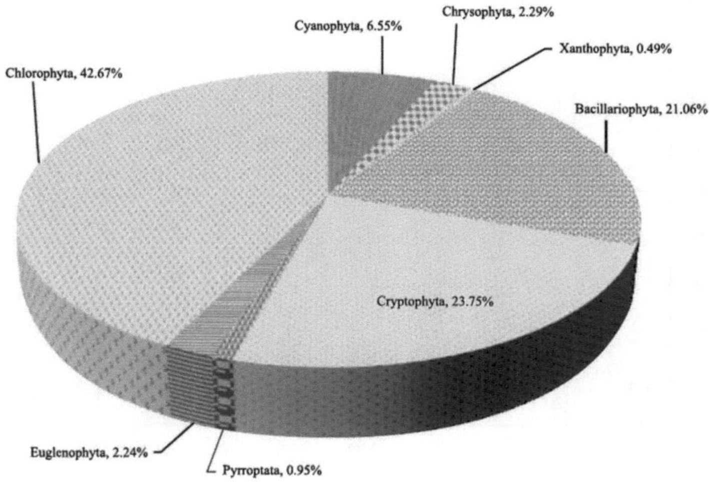

Fig. 9.12 Density proportion of phytoplankton types

Result Analysis

Between April 2010 and February 2011, the quantity characteristic of the phytoplankton types of Dalian Lake was chlorophyta (42.67 %)>cryptophyta (23.75 %)>bacillariophyta (21.06 %)>cyanophyta (6.55 %)>chrysophyta (2.99 %)>euglenophyta (2.24 %)>pyrroptata (0.95 %)>xanthophyta (0.49 %), seen in Fig. 9.12. Clearly, chlorophyta, cryptophyta, bacillariophyta are the main types of phytoplankton. Cyanophyta is evidently not a main type in waters research. On-spot monitoring and experimental data all indicate that there is not the phenomenon of blue-green algae breakout.

Four seasons are divided with "temperature method" in China, namely, it is summer when temperature is more than 22 °C, it is winter when temperature is lower 10 °C, it is spring or autumn when temperature is between the two. It can be from the temperature data (China Statistics Almanac) of each month of 1997–2007 in Shanghai that it is winter between December and February in Shanghai, it is spring between March and May, it is summer between June and September, it is Autumn in October–November.

Between April 2010–February 2011, the density of phytoplankton of Dalian Lake is 4.75×10^6 piece/L on average, fluctuating between $1.99 \times 10^6 \sim 1.56 \times 10^7$ piece/L. Apparently, the highest value of the density of the phytoplankton is in June, the lowest value is in May. The density of the phytoplankton of Dalian Lake retains a relatively low level on the whole. It can be seen from the Fig. 9.13 that the even density of the phytoplankton every month takes on single peak change, the density of phytoplankton in summer is far higher than other three seasons.

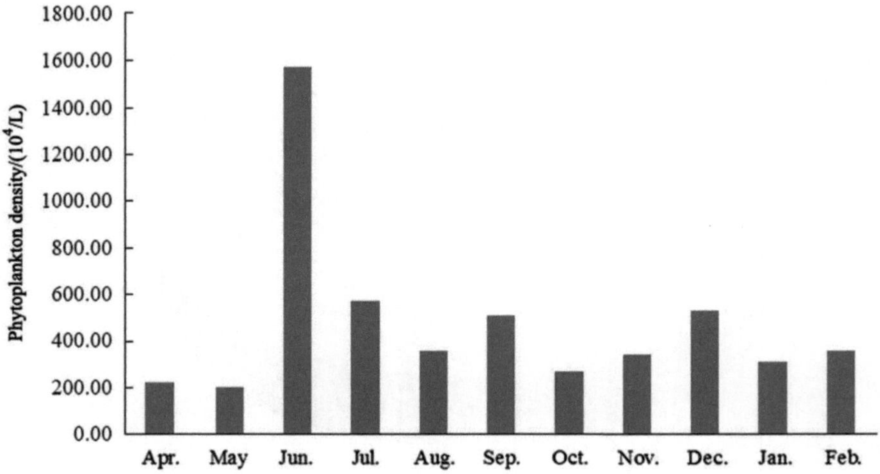

Fig. 9.13 The change trend of the phytoplankton density the water in Dalian Lake

9.3 Other Monitoring Results of Wetland

9.3.1 Monitoring Results from Commissioned Unit

In September 2010, Nanjing University invited and commissioned Qingpu Hydrological Survey Team with relevant qualifications to do water quality monitoring and assessment of project. Nine sampling places (No. 1–9) were set up in the whole project area. Sampling places are seen in Fig. 9.14.

The monitoring results of water quality were showed in Table 9.2. After the project, the water in Dalian Lake is in the state of Class II–III. This shows that the water in Dalian Lake wetland has improved to certain extent after wetland restoration project.

9.3.2 Monitoring Results from Other Units

9.3.2.1 Monitoring by Chinese Academy of Fishery Sciences

After the wetland restoration project, Chinese Academy of Fishery Sciences monitored water quality of project. The sampling time is from September 2009 to April 2010. Five sampling places were set up in the whole area. CK is set up in the fish pond around the project and the other one is set up in the floodgate. The project area has 3 sampling places which were called 1#, 2# and 3#. 1# is near the pond bank, 2# is in the deep water area, 3# is near the quick-infiltration ecological system.

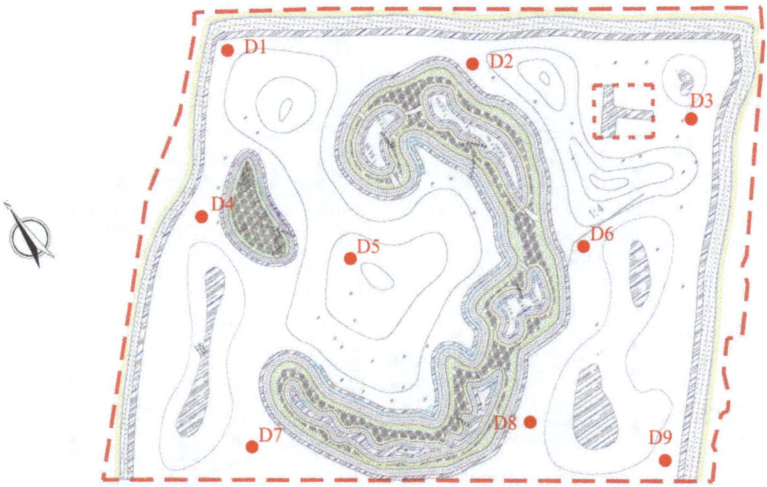

Fig. 9.14 The sampling places of the water quality monitoring

Table 9.2 The monitoring results of water quality

Sampling place	Sampling number	Parameter of water quality					
		pH	COD_{Mn}/ (mg/L)	TP/ (mg/L)	DO/ (mg/L)	NH_4-N/ (mg/L)	TN/ (mg/L)
1	DLH1-1/2/3	7.5	5.2	0.055	4.8	0.21	0.94
2	DLH2-1/2/3	7.5	4.9	0.046	5.0	0.18	0.89
3	DLH3-1/2/3	7.4	5.0	0.043	5.0	0.15	0.76
4	DLH4-1/2/3	7.5	5.2	0.059	4.8	0.05	0.84
5	DLH5-1/2/3	7.6	5.2	0.055	5.0	0.08	0.98
6	DLH6-1/2/3	7.5	4.9	0.047	4.8	0.13	0.89
7	DLH7-1/2/3	7.5	4.9	0.069	5.0	0.13	0.83
8	DLH8-1/2/3	7.5	5.0	0.076	4.9	0.14	0.86
9	DLH9-1/2/3	7.5	5.3	0.071	4.9	0.19	0.83
Average		7.5	5.07	0.05789	4.91	0.14	0.87
Standard	Environment Quality Standard for Surface Water of the People's Republic of China (GB3838—2002)						
Estimate		Non	III	II	Near III	I	III
comments	1. Sampling time is Sep. 4th, 2010						

From Fig. 9.15, we can know that water quality significantly improved after the restoration project. The monitoring results showed that COD, TN and TP in water were decreased by 68, 62 and 74 %. The average NH_4-N is 0.27 mg/L and NO_2-N is 0.02 mg/L. The transparency of water improved significantly.

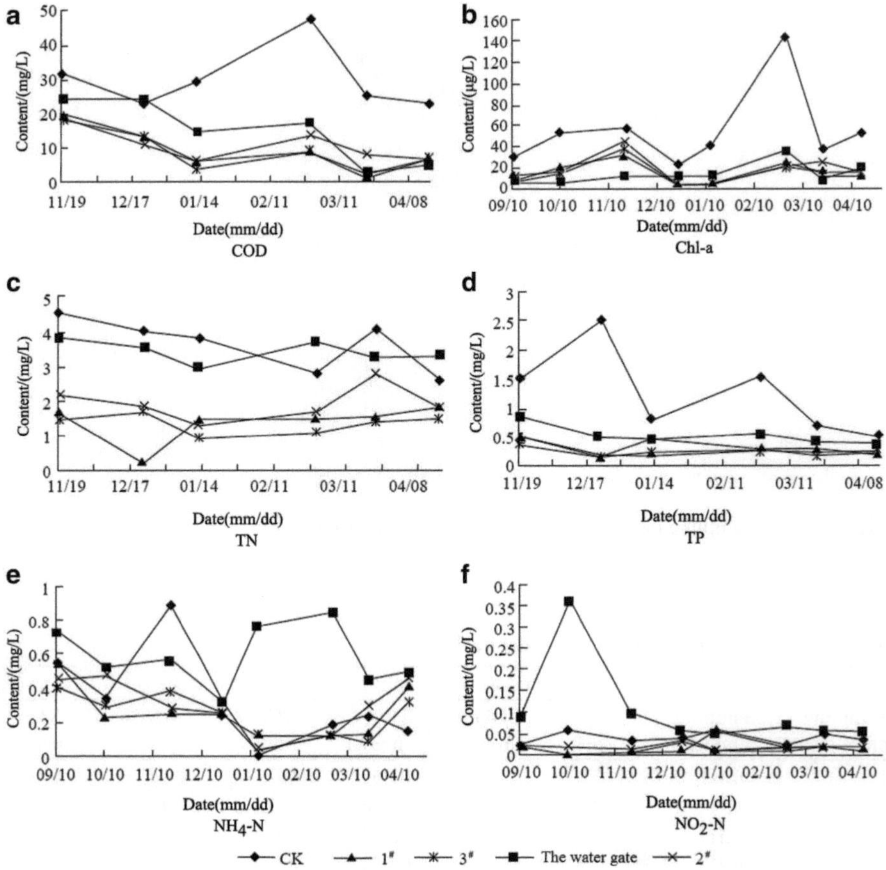

Fig. 9.15 Water quality of the project

9.3.2.2 Monitoring by East China Normal University

From August 2008 to March 2010, East China Normal University monitored the project to assess the effects of the restoration.

Birds

Before the project, totally 65 kinds of birds were found; all of them belong to 11 orders and 27 families. After the project, totally 71 kinds of birds were found; all of them belong to 11 orders and 32 families. There are 11 water birds and ten terrestrial birds which were first found after the project. They are *Bubulcus ibis*, *Botaurus stellaris*, *Anas falcate*, *Anas poecilorhynch, Anas querquedula*, *Anas strepera*, *Aythya nyroca*, *Mergellus albellus*, *Buteo buteo*, *Falco tinnunculus*, *Fulica atra*, *Tringa erythropus*, *Tringa ochropus*, *Ceryle rudi, Hirundo daurica*,

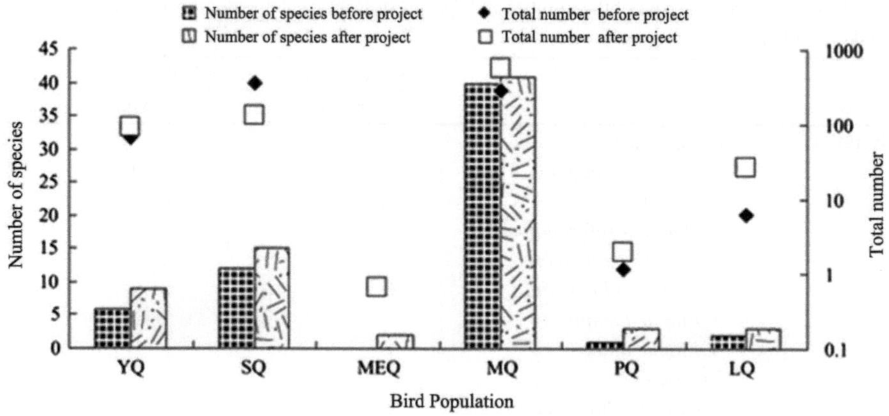

Fig. 9.16 The comparison of birds' species and individual number before and after the demonstration project. *YQ* Swimming birds, *SQ* Waterfowls, *MEQ* Raptor, *MQ* Songbirds, *PQ* Scansores, *LQ* Terrestrial birds

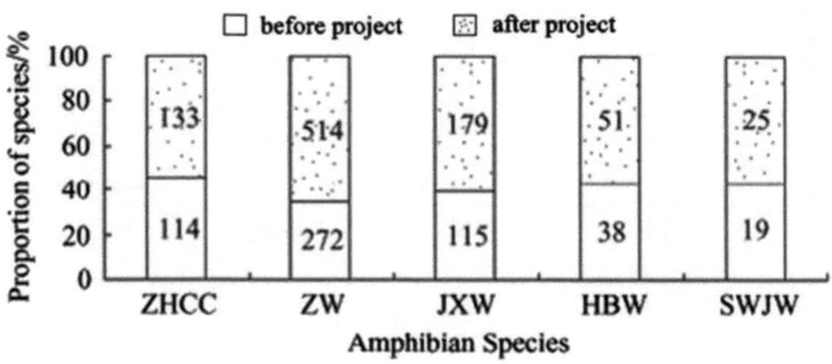

Fig. 9.17 The comparison of amphibian species in the project area before and after the project. *ZHCC* Bufo bufogargarizans, *ZW* Rana limnocharis, *JXW* Rana plancyi, *HBW* Rana nigromaculatta, *SWJW* Microhyla ornata

Lanius sphenocercus, *Acridotheres cristatellus*, *Pica pica*, *Prinia inornata*, *Remiz consobrinus*, *Carduelis sinica*, etc.

In a word, number of species and total number of birds after the project are higher than it before the project (Fig. 9.16).

Amphibians and Reptiles

Before and after the project, we all found five amphibians which included *Bufo bufogargarizans*, *Rana limnocharis*, *Rana plancyi*, *Rana nigromaculatta* and *Microhyla ornate*. Total number of each species after project is higher than before the project (Fig. 9.17).

Table 9.3 The comparison of the number and species of Reptilia in the demonstration area before and after the project

Class	Species	Total number of individual	
		Before the project	After the project
Reptilia	Gekkonidae		
	1. *Gekko japonieus*	1	0
	Colubridae		
	2. *Dinodon rufozonatum*	8	10
	3. *Zaocys dhumnades*	8	8
	4. *Elaphe rufodorsata*	6	6
	5. *Elaphe taeniura*	1	2
	6. *Elaphe dione*	1	1
	Viperidae		
	7. *Gloydius brevicaudus*	2	2
Total	3 families and 7 species	27	29

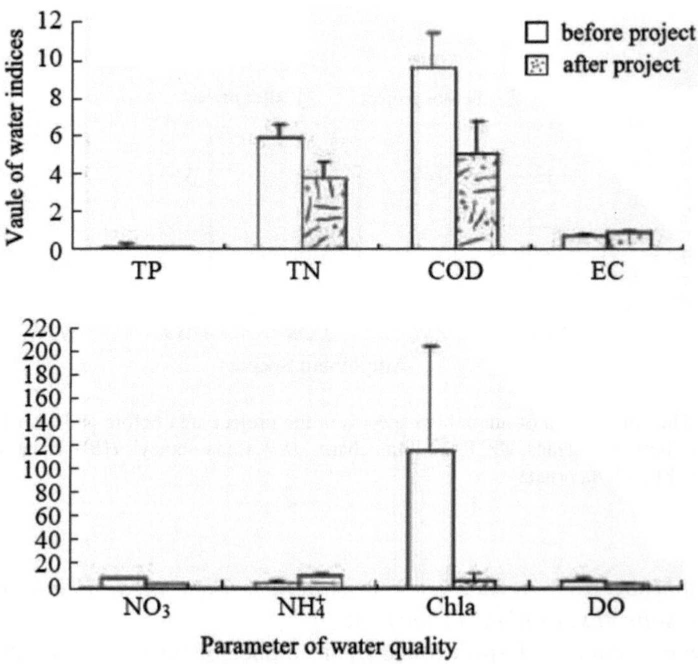

Fig. 9.18 The comparison of eight water quality indices in the demonstration area before and after the project

We also found seven reptiles which included Gekko japonieus, Dinodon rufozonatum, Zaocys dhumnades, Elaphe rufodorsata, Elaphe taeniura, Elaphe dione and Gloydius brevicaudus. Except Gekko japonieus, the total number of reptiles after project is no less than before the project (Table 9.3).

Water Quality

From Fig. 9.18, we can know that after the project, the Chl-a, N, TP, COD, DO in water decreased significantly. This showed that the wetland restoration project had great effects on water quality.

Bibliography

Ge, Z. M., Zhou, X., Wang, K. Y., Chen, L. T., & Wang, T. H. (2009). Ecological planning and benefits analysis on deteriorated lake wetlands: A case study of west suburbs wetland (Shanghai) [J]. *Ecological Economy, 4*, 30–36 (in Chinese).

State Environmental Protection Administration. (2002). *Determination methods for examination of water and wastewater* [M]. Beijing: China Environmental Science Press (in Chinese).

Wu, P. C., Deng, X. J., Lv, H. Z., Peng, S. L., & Zou, J. (2003). Supervision and analysis of water quality in Dongting Lake recovery area [J]. *Journal of Soil and Water Conservation, 17*(1), 134–140 (in Chinese).

Zhu, H., Liu, X. G., Pei, E. L., Guo, W. L., & Xia, S. Z. (2010). Studies on effects of ecological restoration on water quality in Dalian Lake [J]. *Chinese Journal of Environmental Engineering, 4*(8), 1790–1794 (in Chinese).

Chapter 10
Benefit Appraisal of the Wetland

10.1 The Ecological and Environmental Benefit of the Project

10.1.1 Ecological Benefit

10.1.1.1 Improvement of Biodiversity

The construction of the project has restored 60 kinds of aquatic plants and organisms of the wetland, such as arbors, shallow water emergent plants, deep water emergent plants, floating leaved plants, submerged plants, etc., which taking on the biodiversity of plant ecosystem after construction that changes with the characteristics of terrain. The wetland after restoration has been characterized with lush aquatic plants and organisms, plenty and limpid water, ecological stability and the stagger of all species that will not only produce biodiversity but also have good water-purifying effect all the year round.

Shore wetland belt is the belt with most diversified aquatic organisms. Wetland protection and restoration will provide a good habitat for aquatic organisms where existing species can be effective protected and where other species can be introduced from Yangtze River and other near rivers. The ecological continuity of lakeshore provides another kind of buffering mechanism for guaranteeing the ecological stability of lakes. When flood or water pollution occurs, shoal wetland will provide shelter for submerged plants and the aquatic animals in shallow water; when the flood or water pollution passes, these organisms can gradually occupy their original territories and original benign ecology can be restored.

S. An and L. Wang, *Wetland Restoration: Shanghai Dalian Lake Project*, Springer Environmental Science and Engineering, DOI 10.1007/978-3-642-54230-5_10,

10.1.1.2 Improvement of Biological Indexes

Aquatic plants and animals will also have benign development with the change of the ecological environment of waters while wetland plant ecosystem improves. The project area will undergo self-sustaining development after human assistance stops.

10.1.1.3 Increase of Ecological Asset Value

In contrast of traditional high yield fishponds and wetland, the former contributes nothing in waste disposal, biodiversity protection and recreative and cultural value; but the latter can respectively yields RMB 16,050, 2,210, 4,910/(hectare·year) in ecological services value; and 625 mu (41.7 ha) model project can provide RMB 966,190/year in ecological services value.

10.1.1.4 Carbon Dioxide Emission Reduction

A waste water treatment plant processes 1 t waste water while releases 0.30 kg (equivalent to carbon 0.08 kg) carbon dioxide, but the wetland treats the same amount of waste water while only emits 0.04 kg carbon (carbon dioxide and methane). From this, the project treats 868,700–956,300 t waste water a year while can reduce carbon emission 34,748–38,252 kg (equivalent to 127–140 t carbon dioxide emission).

10.1.2 Environmental Benefit

10.1.2.1 Improvement of Physical Indexes

Removal of Sestons

Researches testify that reed and cattail all can notably fall the sestons in water, up to 50–80 % in contrast mainly through absorbing sestons. But the two kinds of plants have not noticeable difference. In the project, consecutive reed belt can employ its own branches and communities with certain density to create static water environment to intercept and absorb suspended particles in water.

Improvement of the Aesthetic Quality of Water

The improvement of the aesthetic quality of water mainly embodies on water transparency and color. Of them, the transparency is 80 cm and an important index of landscape water. The seston tends to be the only factor to determine transparency in non-bloom period. Thus, removing the sestons out of water will obviously improve the transparency of water. As the emergent plant project was carried out, especially, after the project construction ended, the transparency of the water in the project area has been greatly improved.

Water color is mainly affected by colored organic groups, iron oxide, sulfide, etc. The water in estuaries generally does not generate anaerobic environment, so sulfide will not occur. In this, colored organic groups probably is the main factor to determine water color. Aquatic plants greatly removed these materials after the operation of the project, and water color meets the sensory need of people.

10.1.2.2 Improvement Effect of Chemical Indexes

Lots of aquatic plants planted in the project area have relatively strong ability of removing pollutants in water and bottom sludge. According to researches, Zizania aquatica owns the ability of purifying sestons, chloride, nitride, sulfate in water, with the rate of removing nitrogen and phosphor up to 227.1, 28.9 $g/(hm^2 \cdot year)$ respectively, and COD of phytoplanktons with relatively great amount of organisms up to 63.82 %; and scirpus tabernaemontani, cattail, pondweed can remove 95 % of organic pollutants.

Aquatic plants absorb a little of organics, heavy metal and nutrients, etc., through absorption or decomposition. The researches of floating-leaved plant trapa incisa indicate that if water stays at trapa incisa communities for 10 day, the COD_{Mn} in it can lower 80 %. Reed can absorb mercury and lead; phytoplankton often enrich pollutants around their roots; nymphoides has certain absorbing and purifying role of heavy metal, such as Cd, As, Hg, Zn, etc. Of them, enriching multiple of Cu is 880, Cd 175, Cr 2046.

Removing nutrients is the most important chemical index required for wetland. The emergent aquatic plants in wetland system remove TN mainly through the absorption of plant tissues and rhizospheric microorganism, roots stranded, the nitration and anti-nitration around the roots. Wetland types, plant types, climate, temperature, water staying time all have affection on removing TN. Wetland plants can strengthen the activities of the microorganisms in filling and improve the absorption of rhizospheric microorganism, roots stranded, the nitration and anti-nitration around the roots; therefore, human nitrogen-removing effect is enhanced.

Lots of shallow water emergent plants are planted for the wetland project, such as Thalia debata, water fennel, rhizoma, acori graminei, willow herb, etc. They all can absorb nutrients in water and bottom sludge, such as, nitrogen, phosphor, etc.; and even can restrict the reproduction of algae.

10.2 Social and Economic Benefit of the Project

The project belongs to the ecological wetland construction project with social public welfare nature. It has the incalculable social benefit of improving water environment, building healthy water ecosystem, increasing biodiversity and strengthening

Table 10.1 Profit analysis of organic vegetable planted

Profit	Zizania aquatica	Water fennel	Arrowhead	Chufa	Water spinach
Area/mu	1.66	1.66	0.55	0.55	1.11
Yield/(mu/kg)	3,321.8	6,643.5	442.9	553.6	2,214.5
Market price/(yuan/kg)	3	3	2.4	4	2
Profit/yuan	16,538.6	33,084.6	584.6	1,217.9	4,916.2
Total profit/yuan	56,341.9 (cost 25,000)				
Net profit/yuan	31,341.9				

surrounding residents' awareness of wetland environment protection, etc. Its main representations are as follow:

1. The rivers surrounding Dalian Lake and existing fishponds are interconnected through river system construction system. Water flow and diluting ability are strengthened in accordance with specific condition and associated with the construction of water and soil conservation, greening and ecological river landscape. Water self-purifying ability is intensified by the implementation of river and wetland engineering to conserve water and restore organisms. In addition, the water from upper reaches can be purified and filtered through wetland construction, so the water environment of lower reaches and the local area can be improved. Furthermore, 868,700–956,300 t water up to Class III surface water is provided for Shanghai each year, of them, rural domestic sewage that is treated and reaches standards plus wetland water that is purified 10,950–14,600 t and the water from wetland treatment system 857,750–941,700 t.
2. The project has improved local flood fighting ability, created a more safe and stable productive and life environment for local people. Therefore, more investment can be attracted to promote local development. There is great potential of value increase in local land, houses and fixed assets under market economy. This can enhance local construction speed to ensure sustainable and stable development of local economy.
3. The project has provided employment for 4–10 people in the course of its implementation, management and long-term operation, made 800 people in Lianhu village, 20,000 people in Jinze, Liantang and Zhujiajiao towns and 200 million people in Qingpu has the awareness of water source protection, and changed the life habit of 50,000 residents in the communities near water source, actively participating water source protection.

In addition, the project area still contains certain practical economic value in economic planting and breeding. To be specific, they are:

The project area accounts for over 160 mu waters after transformation, whose aquatic production is 1/5 of the yield of the original fishponds estimated in accordance with the principle of seedling ratio and biomass amount, 1/15–1/10 of the original yield estimated in accordance with the use of logistic growth curve in animal community biomass amount and whose profit is RMB 300–500 per mu and 48,000–80,000 per year estimated in accordance with local original profit.

The profit of aquatic vegetables, such as zizania aquatica, arrowhead, etc. That are planted in the project area is seen in Table 10.1.

Bibliography

Armin, W., Lorenz, S. C., & Jahnig, D. H. (2009). Remaindering German low land streams: qualitative and quantitative effects of restoration measures on hydromorphology and macroinvertebrates [J]. *Environmental Management, 44*(4), 745–754.

Huo, Y. Z., He, W. H., Luo, K., Wang, Y. Y., Zhang, Y. J., Tian, Q. T., & He, P. M. (2010). Bioremediation efficiency of applying Daphnia magna and submerged plants: a case study in Dishui Lake of Shanghai, China [J]. *Chinese Journal of Applied Ecology, 21*(2), 495–499 (in Chinese).

Pei, E. L., Guo, W. L., & Xia, S. Z. (2010). Distribution and dynamics of planktons in ecological restoration zone of Dalian Lake, Shanghai [J]. *Chinese Journal of Wildlife, 31*(3), 154–156 (in Chinese).

Xu, D. F., Xu, J. M., Wang, H. S., Luo, A. C., Xie, D. C., & Ying, Q. S. (2005). Absorbability of wetland plants on N and P from eutrophic water [J]. *Plant Nutrition and Fertilizer Science, 11*(5), 597–601 (in Chinese).

Appendices

Appendix 1

Table 1.1 Types and distribution of the phytoplankton of Dalian Lake

Phylum	Genus	Species	I	II	III	IV	V	VI	VII	VIII	IX	X
Cyanophyta												
	Oscillatoria	*O.* sp.				+						
	Spirulinas	*S. major*				+			+		+	
	Phormidium	*P.* sp.	+	+	+	+	+	+	+	+		
	Anabaena	*A.* sp.		+	+	+					+	
		A. oscillarioides		+						+	+	
	Lyngbya	*L. limnetica*							+			
	Chroococcuss	*C. minutus*	+	+	+	+	+			+		
		C. minor	+	+								+
	Coelosphaerium	*C. dubium*								+		
	Merismopedia	*M. tenuissima*	+	+	+		+	+	+	+	+	+
		M. glauca		+	+	+						
		M. elegans	+	+	+	+	+	+				
	Gloeocapsa	*G. punctata*		+								
	Dactyloccocopsis	*D. rhaphidioide*			+							+
		D. rhaphidioide f. falciformis		+	+	+	+		+	+		+
		D. acicularis							+			
	Raphidiopsis	*R. sinensia*		+	+							
Xanthophyta												
	Tribune	*T.* sp.			+		+		+			
Pyrrophyta												
	Peridinium	*P.* sp.			+							

(continued)

S. An and L. Wang, *Wetland Restoration: Shanghai Dalian Lake Project*, Springer Environmental Science and Engineering, DOI 10.1007/978-3-642-54230-5,

Table 1.1 (continued)

Phylum	Genus	Species	I	II	III	IV	V	VI	VII	VIII	IX	X
Cryptophyta												
	Cryptomonas	*C. ovata*	+	+		+	+	+			+	
		C. erosa		+	+	+	+	+		+	+	+
Euglenophyta												
	Euglena	*E. acus*		+		+						
		E.geniculat			+							
	Phacus	*P. longicauda*		+	+						+	+
		P. suecicus										+
		P. inflexus										+
		P. tortus										+
Bacillariophyta												
	Synedra	*S. acus*			+	+	+	+				+
		S. ulna	+	+	+	+	+	+	+		+	
		S. ulna var. contracta				+						
	Cyclotella	*C. meneghiniana*	+	+			+		+	+	+	+
	Cymbella	*C. centricosa*				+						
	Gomphonema	*G. constrictum*				+						
	Navicula	*N. eymei*	+			+	+	+	+			+
	Melosira	*M. granulate*								+	+	
		M. varians	+	+	+		+					
	Nitzschia	*N.* sp.				+	+		+	+		
		N. longissima			+	+						
		N. lorenziana		+	+	+	+					
Chlorophyta												
	Chlamydomona	*C. debaryana*										+
		C. simplex	+	+	+		+	+		+	+	
		C. ovalis		+	+		+			+		+
		C. mutabilis		+	+							+
		C. braunii								+		+
	Schroederia	*S. spiralis*		+	+							+
		S. nitzschioides	+	+			+	+			+	
	Coelastrum	*C. microporum*		+	+		+					
	Selenastrum	*S. minutum*		+	+		+	+		+	+	
	Staurastrum	*S* .sp.			+							
	Tetraëdron	*T. regulare*			+							
		T. caudatum			+							
		T. trigonum		+	+	+	+			+	+	+
	Palmwllococcus	*P. miniatus*										

(continued)

Table 1.1 (continued)

Phylum	Genus	Species	I	II	III	IV	V	VI	VII	VIII	IX	X
	Crucigenia	*C. apiculata*					+	+				
		C. lauterbornei		+						+		
		C. quadrata	+	+		+	+			+		+
		C. tetrapedia		+	+		+	+				
		C. ellipsoedea										
	Chlorella	*C. vulgaris*		+	+	+	+	+	+	+	+	+
		C. ellipsoedea										+
	Micractinium	*M. pusillum*								+		+
	Pteromonas	*P. angulosa*			+							+
		P. aculeata		+								
	Oocystis	*O.* sp.						+			+	
	Quadrigula	*Q .chodatii*		+								+
	Scenedesmus	*S. acuminatus*		+	+					+		+
		S. dimorphus	+	+	+		+					+
		S. quadricauda	+	+	+		+	+	+	+	+	+
		S. arcuatus	+	+								
	Ulothrix	*Ulothrix sp.*				+	+			+		
	Tetrastrum	*Tetrastrum hastiferum*					+			+		
	Pediastrum	*P. simplex var. duodenarium*	+			+						
		P. boryanum					+					
		P. tetras			+							
		P. biradiatum								+		
	Ankistrodesmus	*A. falcatus*		+		+	+	+		+	+	
		A. falcatus var. *mirabilis*	+	+	+							
		A. sangustus									+	
		A. acicularis					+				+	+
	Chlorogonium	*C. elongatum*		+			+					
	Westella	*W. botryoides*	+	+			+				+	+
	Chodetella	*C.* sp.										+
	Actinastrum	*A. hantzschii*		+		+	+			+	+	+
	Dictyosphaerium	*D. ehrenbergianum*										+
	Eudorina	*E. elegans*		+	+	+	+	+				
	Pandorina	*P. morum*				+						
Total: 4 Phylum	48 Genus	86 Species	19	45	39	30	36	18	13	28	22	32

Table 1.2 Phytoplankton density of Dalian Lake

	Sample place I	Sample place II	Sample place III	Sample place IV	Sample place V	Sample place VI	Sample place VII	Sample place VIII	Sample place IX	Sample place X	Total	Average
Cyanophyta	3,877,419	37,548,387	11,777,126	529,750	58,554,839	622,500	118,250	16,116,129	4,187,097	22,761,290	156,092,787	15,609,279
Xanthophyta	0	0	269,795	0	129,032	0	750	0	0	0	399,577	39,958
Pyrrophyta	0	0	234,604	0	0	0	0	0	0	0	234,604	23,460
Cryptophyta	12,903	1,032,258	797,654	2000	670,968	200	0	335,484	225,806	658,065	3,735,338	373,534
Euglenophyta	0	38,710	82,111	250	0	0	0	0	25,806	245,161	392,038	39,204
Bacillariophyta	174,194	683,871	1,771,261	20,500	1,419,355	5250	149,500	2,503,226	335,484	154,839	7,217,479	721,748
	335,484	6,038,710	3,577,713	65,500	6,567,742	13,000	21,000	7,793,548	3,974,194	7,341,935	35,728,825	3,572,883
Chlorophyta	4,400,000	45,341,936	18,510,264	618,000	67,341,936	640,950	289,500	26,748,387	8,748,387	31,161,290	203,800,650	20,380,065

Table 1.3 Types and distribution of the zoobenthos in Dalian Lake

Phylum	Genus	Species	I	II	III	IV	V	VI	VII	VIII	IX	X
Rotifera												
	Brachionus	*B. calyciflorus*	+	+	+	+						+
		B. angularis	+		+	+						+
		B. forficula					+					
		B. falcatus	+									
		B. urceus										+
		B. diversicornis	+	+		+	+					
	Keratella	*K. cochlearis*	+									
		K. quadrata					+					
		K. valga										
	Polyarthra	*P. trigla*			+							
	Filinia	*F. maior*		+		+	+					
	Asplanchna	*A. sp.*	+	+	+	+	+					+
	Epiphanes	*E. senta*				+						
Cladocera												
		Nauplii	+		+	+	+	+	+			
	Bosmina	*B. longirostris*	+	+		+		+	+			
		B. fatalis	+			+	+	+	+			
	Diaphanosoma	*D. brachyurum*	+	+		+	+	+		+		+
	Moina	*M. affinis*	+			+	+	+	+			+
	Alona	*M. intermedia*					+		+			
Copepoda												
	Sinocalanus	*S. sinensis*	+			+	+		+	+	+	+
	Schmackeria	*S. forbesi*				+	+	+	+	+		
	Sinodiaptoms	*S. sarsi*	+			+	+	+		+		+
	Cyclops	*C. vicinus*	+	+	+	+	+	+	+	+	+	+
	Mesocyclops	*M. leuckarti*	+			+	+	+	+		+	+
Total	15 Genus	24 Species	14	7	6	17	15	8	8	5	3	10

Appendix 2

Attachment 2.1. Environment Quality Standard for Surface Water of the People's Republic of China

[Title]: Environment Quality Standard for Surface Water of the People's Republic of China
[Issued by]: State Environmental Protection Administration
[Serial No.]: GB3838—2002
[Date of issuing]: 2002-4-28
[Date of implementation]: 2002-6-1

State Environmental Protection Administration Public Announcement for Environment Quality Standard for Surface Water

Environment Quality Standard for Surface Water is now approved as national environment quality standard in order to implement *Environment Protection Law and Water Pollution Prevention and Control Law*, strengthen surface water environment management, prevent and control environment pollution and protect human health and jointly issued by *State Environmental Protection Administration and State Administration for Quality Supervision and Inspection and Quarantine*.

Its name and serial No. are as follow:

Environment Quality Standard for Surface Water (GB3838—2002).

The standard is published by China Environmental Science Press and will be implemented on June 1, 2002.

The public announcement is hereby specially made.

State Environmental Protection Administration
April 26, 2002

Environment Quality Standard for Surface Water of the People's Republic of China

Issued by State Environmental Protection Administration and State Administration for Quality Supervision and Inspection and Quarantine

Preface

This standard is enacted and made in order to implement *Environment Protection Law of the People's Republic of China* and *Water Pollution Prevention and Control Law of the People's Republic of China*, prevent and control water pollution, protect surface water quality, safeguard human health and maintain good ecological system.

The items of the standard are divided into basic item of environment quality standard for surface water, supplementary item of the surface water resource places of concentrated drinking water and special item of the surface water resource places of concentrated drinking water. The basic item of environment quality standard for surface water applies to the surface waters with use functions in China, such as rivers, lakes, channels, ditches, reservoirs, etc.; supplementary item and special item of the surface water resource places of concentrated drinking water applies to Class I reserves and Class II reserve of the surface water resource places of concentrated drinking water. The special item of the surface water resource places of concentrated drinking water is chosen by the administrative competent department of environment protection of local people's government at or above county level in accordance with local surface water characteristics and the requirements of environmental management; The supplementary item of the surface water resource places of concentrated drinking water and the special item chosen and determined are taken as the supplementary indexes of the basic item.

The items of the standard contains 109 indexes altogether. Of them, the basic item of environment quality standard for surface water includes 24 indexes, the supplementary item of the surface water resource places of concentrated drinking water contains 5 indexes, and the special item of the surface water resource places of concentrated drinking water contains 80 indexes.

Compared with GHZB1—1999, the standard adds one index to the basic item of environment quality standard for surface water, deletes three indexes, basic requirements and nitrite, nonionic ammonia and KTN, adjusts sulfate, chloride, nitrate, iron, manganese into the supplementary item of the surface water resource places of concentrated drinking water, revises the values of 7 indexes pH, dissolved oxygen, ammonia nitrogen, total phosphor, permanganate index, led, fecal coliform, adds forty indexes to the special item of the surface water resource places of concentrated drinking water.

The administrative competent department of environment protection and related department of local people's government at or above county level supervise and manage all kinds of water of surface water pursuant to the standard and duties and responsibilities and division of labor.

The estuary waters of the surface water connected to offshore waters shall be managed in accordance with water environment functions and the corresponding type standard values of the standard. The waters of offshore water function area shall be managed pursuant to functions of use and the corresponding type standard values of *Seawater Quality Standard*. The single fishing waters that are approved and determined shall be managed in accordance with *Seawater Quality Standard*. The water quality of urban treated sewage and the industrial waste water near to urban sewage that are used for agricultural irrigation water shall be managed in accordance with *Standards for Irrigation Water Quality*.

Environment Quality Standard for Surface Water (GB 3838—83) was issued for the first time, revised in 1988 for the first time, in 1999 for the second time, this time for the third time. *Environment Quality Standard for Surface Water* (GB3838-2002) will be implemented since June 1, 2002; *Environment Quality Standard for Surface Water* (GB 3838—88) and *Environment Quality Standard for Surface Water* (GHZB1-1999) will be abolished at the same time.

The standard is proposed and supervised State Environmental Protection Administration Science and Technology and Standard Department.
The standard is revised by China Academy of Environment Science.
The standard is approved by State Environmental Protection Administration on April 26, 2002.
The standard is interpreted by State Environmental Protection Administration.

Environment Quality Standard for Surface Water

1. Scope

1.1. The standard, in accordance with the environmental function classification and protective goals of surface water, stipulates the items and limits that shall be controlled in water environment quality, water quality appraisal, the analysis methods of water quality items and the implementation and supervision.

1.2. The standard applies to the surface waters with use functions in the territory of the People's Republic of China, such as, rivers, lakes, channels, ditches, reservoirs, etc. For the waters with special functions, the corresponding standards of the water quality of special water shall be exercised.

2. Quoted Standards

The clauses in Hygienic Specifications for Drinking Water (Ministry of Health, 2001) and the analysis methods listed in the Tables 2.4, 2.5, and 2.6 of the standard, if they are quoted in the standard, form the clauses of the standard and is effective the same with the standard. The latest edition of the above mentioned standard and specifications shall be used when they are revised.

3. The Waters Functions and Standard Classification

The waters functions are divided into five classes in accordance with the environment functions and protective goals.

Class I mainly applies to water source and national natural reserves;

Class II mainly applies to the Class I reserves of the surface water resource places of concentrated drinking water, the habitats of rare aquatic organisms, the spawning grounds of fish and shrimps and the like, the feeding grounds of young fish, etc.;

Class III mainly applies to the Class II reserves of the surface water resource places of concentrated drinking water, the wintering ground of fish and shrimps, migrating channels, breeding areas of aquatic products and other fishing waters and swimming areas;

Class IV mainly applies to common industrial water areas and the water areas for recreation that human body does not directly touch;

Class V mainly applies to agricultural water areas and the waters for common landscape.

The standards of the basic item of the environment quality standard for surface water is divided into five classes pursuant to the above the five classes of surface waters functions. For different functions and classes, the standard values of corresponding classes are exercised. The standards of high classes of waters functions are stricter that those of low classes of water functions. For the same waters with diversified functions, the corresponding standard values of the highest class of waters functions shall be exercised. Realizing water functions is of the same meaning with reaching the classes of functions.

4. Standard Values

4.1 The standard limits of the basic item of environment quality standard for surface water are seen in appendix Table 2.1.
4.2 The standard limits of the supplementary item of the surface water resource places of concentrated drinking water are seen in appendix Table 2.2.
4.3 The standard limits of the special item of the surface water resource places of concentrated drinking water are seen in appendix Table 2.3.

5. Water Quality Appraisal

5.1 For the environment quality appraisal of surface water, single factor appraisal shall be carried out in accordance with the classes of water functions that should be realized and choosing the standards of corresponding classes. The appraisal result should states how water quality reaches standards; if water quality exceeds standards, the indexes of exceeding standards and the multiple of exceeding standards shall be stated.
5.2 For the water with obvious high, level and dry water periods, water appraisal shall be conducted by water periods.
5.3 The indexes of the water quality appraisal of the surface water resource places of concentrated drinking water shall contain the basic indexes in appendix Table 2.1, the supplementary item in appendix Table 2.2 and the special item that the administrative competent department of local people's government at or above county level chooses from appendix Table 2.3 and determines.

6. Water Quality Monitoring

6.1 The standard values stipulated in the standard requires that sample water sinks for 30 min naturally after it is collected and the upper non-sink part shall be collected in accordance specified methods.
6.2 The sample places for surface water quality monitoring and monitoring frequency shall be in accordance with the requirements of national technical specifications for the environment monitoring of surface water.
6.3 The analysis methods of the water quality items of the standard shall adopt the ones specified in appendix Table 2.4, 2.5, and 2.6 with priority, also can adopt ISO method system and other analysis methods with equivalent effect but it is necessary to conduct applicable tests.

Table 2.1 The standard limits of the basic items of the environment quality standard for surface water (Unit: mg/L)

S/N	Class	Class I	Class II	Class III	Class IV	Class V
1	Water temperature (°C)	Artificial change of ambient temperature shall be limited to Weekly even maximum temperature rise ≤1 Weekly even maximum temperature fall ≤2				
2	pH value (non-dimensional)	6–9				
3	Dissolved oxygen ≥	Saturation rate 90 %(or 7.5)	6	5	3	2
4	Permanganate index	2	4	6	10	15
5	Chemical oxygen demand (COD)≤	15	15	20	30	40
6	Five-day BOD≤	3	3	4	6	10
7	Ammonia nitrogen (NH_3-N)≤	0.15	0.5	1.0	1.5	2.0
8	Total phosphor (on the basis of P)≤	0.02 (Lake, Reservoir 0.01)	0.1 (Lake, Reservoir 0.025)	0.2 (Lake, Reservoir 0.05)	0.3 (Lake, Reservoir 0.1)	0.4 (Lake, Reservoir 0.2)
9	Lake and reservoir, total phosphor (on the basis of N)≤	0.2	0.5	1.0	105	2.0
10	Copper≤	0.01	1.0	1.0	1.0	1.0
11	Zinc≤	0.05	1.0	1.0	2.0	2.0
12	Fluoride (on the basis of F^-) ≤	1.0	1.0	1.0	1.5	1.5
13	Selenium≤	0.01	0.01	0.01	0.02	0.02
14	Arsenic≤	0.05	0.05	0.05	0.1	0.1
15	Mercury≤	0.00005	0.00005	0.0001	0.001	0.001
16	Cadmium ≤	0.001	0.005	0.005	0.005	0.01
17	Chromium (six price)≤	0.01	0.05	0.05	0.05	0.1
18	Lead≤	0.01	0.01	0.05	0.05	0.1
19	Cyanide≤	0.005	0.05	0.2	0.2	0.2
20	Volatile phenol≤	0.002	0.002	0.005	0.01	0.1
21	Petroleum≤	0.05	0.05	0.05	0.5	1.0
22	Anionic surfactant≤	0.2	0.2	0.2	0.3	0.3
23	Sulfide≤	0.05	0.1	0.2	0.5	1.0
24	Fecal Coliform (piece/L)≤	200	2,000	10,000	20,000	40,000

Table 2.2 The standard limits of the basic item of environment quality standard for surface water (Unit: mg/L)

S/N	Item	Standard
1	Sulfate (on the basis of SO_4^{2-})	250
2	Cyanide (on the basis of Cl^-)	250
3	Nitrate (on the basis of N)	10
4	Iron	0.3
5	Manganese	0.1

Table 2.3 The standard limits of the special item of environment quality standard for surface water (Unit: mg/L)

S/N	Item	Standard value	S/N	Item	Standard value
1	Trichloromethane	0.06	11	Tetrachloroethylene	0.04
2	Carbon tetrachloride	0.002	12	Chloroprene	0.002
3	Tribromethane	0.1	13	Hexachlorobutadiene	0.0006
4	Dichloromethane	0.02	14	Styrene	0.02
5	1,2- dichloroethane	0.03	15	Formaldehyde	0.9
6	Epoxy chloropropane	0.02	16	Acetaldehyde	0.05
7	Chloroethylene	0.005	17	Acrolein	0.1
8	1,1-dichloroethylene	0.03	18	Grasex	0.01
9	1,2-dichloroethylene	0.05	19	Benzene	0.01
10	Triclene; circosolv	0.07	20	Toluene	0.7
21	Ethylbenzene	0.3	51	Active chlorine	0.01
22	①Dimethylbenzene	0.5	52	DDT	0.001
23	Cumin	0.25	53	Lindane	0.002
24	Chlorobenzene	0.3	54	Heptachlor epoxide	0.0002
25	1,2-dichlorobenzene	1.0	55	Parathion	0.003
26	1,4-dichlorobenzene	0.3	56	Parathion-methyl	0.002
27	②Trichlorobenzene	0.02	57	Malathion	0.05
28	③Tetrachlorobenzene	0.02	58	Dimethoate	0.08
29	Hexachlorobenzene	0.05	59	DDVP	0.05
30	Nitrobenzene	0.017	60	Dipterex	0.05
31	④Dinitro benzene	0.5	61	Systox	0.03
32	2,4- DNT	0.0003	62	Chlorothalonil	0.01
33	2,4,6- trinitrotoluene	0.5	63	Carbaryl	0.05
34	⑤eNitrochlorobenzene	0.05	64	Deltamethrin	0.02
35	2,4-dinitrochlorobenzene	0.5	65	Atrazine	0.003
36	2,4-dichlorophenol	0.093	66	Benzene and (a) pyrene	2.8×10^{-6}
37	2,4,6-trichlorophenol	0.2	67	Methylmercury	1.0×10^{-6}
38	Pentachlorophenol	0.009	68	⑥fPolychlorinated biphenyl	2.0×10^{-6}
39	Phenylamine	0.1	69	Microcystin-LR	0.001
40	Benzidine	0.0002	70	Yellow phosphorus	0.003
41	Acrylic amide	0.0005	71	Molybdenum	0.07
42	Acrylonitrile	0.1	72	Cobalt	1.0
43	Dibutyl phthalate	0.003	73	Beryllium	0.002
44	Bisphthalate (2-cthylhexyl)	0.008	74	Boron	0.5

(continued)

Table 2.3 (continued)

S/N	Item	Standard value	S/N	Item	Standard value
45	Hydrazine hydrate	0.01	75	Antimony	0.005
46	Tetraethyllead	0.0001	76	Nickel	0.02
47	Pyridine	0.2	77	Barium	0.7
48	Turpentine	0.2	78	Vanadium	0.05
49	Picric acid	0.5	79	Titanium	0.1
50	n-butyl xanthate	0.005	80	Thallium	0.0001

Note:
①Dimethylbenzene: Xylene, Refers to metaxylene, Neighbors-xylene.
②Trichlorobenzene: Refers to 1,2,3- trichlorobenzene, 1,2,4- trichlorobenzene,1,3,5- trichlorobenzene.
③Tetrachlorobenzene: Refers to1,2,3,4-etrachlorobenzene, 1,2,3,5- tetrachlorobenzene, 1,2,4,5-tetrachlorobenzene.
④Dinitro benzene: Refers to dinitro benzene, nitrobenzene, Neighbors-two of nitrobenzene.
⑤Nitrochlorobenzene: Refers to nitrochlorobenzene, -the chlorobenzene between, Neighbors-the nitro chlorobenzene.
⑥Polychlorinated biphenyl: Refers to PCB-1016, PCB-1221, PCB1232, PCB1242, PCB-1248, PCB-1254, PCB-1260.

Table 2.4 Analysis method for basic items under environmental quality standard of underground water

No.	Basic item	Analysis method	Lower limit/ (mg/L)	Method origin
1	Water temperature	Thermometer method		GB 13195—91
2	pH	Glass electrode method		GB 6920—86
3	Dissolved oxygen	Iodometric method	0.2	GB 7489—89
		Electrochemistry probe method		GB 11913—89
4	Permanganate index		0.5	GB 11892—89
5	Chemical oxygen demand	Dichromate titration	5	CB 11914—89
6	5-day biochemical oxygen demand	Attenuation & inoculate method	2	GB 7488—87
7	Ammonia nitrogen	Nessler reagent colorimetry	0.05	GB 7479—87
		Salicylic acid spectrophotometry	0.01	GB 7481—87
8	Total phosphorus	Ammonium molybdate spectrophotometry	0.01	GB 11893—89
9	Total nitrogen	Alkaline potassium persulfate digestion UV spectrophotometric method	0.05	GB 11894—89
10	Copper	2,9-dimethyl-1,10-phenanthroline spectrophotometry	0.06	GB 7473—87
		Sodium diethyldithiocarbamate spectrophotometry	0.010	GB 7474—87
		Atomic absorption spectrophotometry (integrated extraction method)	0.001	GB 7475—87
11	Zinc	Atomic absorption spectrophotometry	0.05	GB 7475—87

(continued)

Table 2.4 (continued)

No.	Basic item	Analysis method	Lower limit/ (mg/L)	Method origin
12	Fluoride	Fluorine reagents spectrophotometry	0.05	GB 7483—87
		Ion selective electrode method	0.05	GB 7484—87
		Ion chromatography	0.02	HJ/T 84—2001
13	Selenium	2,3-diaminonaphthalene fluorescence method	0.00025	GB 11902—89
		Graphite furnace atomic absorption spectrophotometry	0.003	GB/T 15505—1995
14	Arsenic	Silver diethyldithiocarbamate spectrophotometric method	0.007	GB 7485—87
		Cold atomic fluorescence method	0.00006	[a]
15	Mercury	Cold atomic absorption spectrophotometry	0.00005	GB 7468—87
		Cold atomic fluorescence method	0.00005	[a]
16	Cadmium	Atomic absorption spectrophotometry (integrated extraction method)	0.001	GB 7475—87
17	Chromium (hexavalent chromium)	Diphenylcarbohydrazide spectrophotometric method	0.004	GB 7467—87
18	Lead	Atomic absorption spectrophotometry (integrated extraction method)	0.01	GB 7475—87
19	Total cyanide	Isonicotinic acid-pyrazolone colorimetry	0.004	GB 7487—87
		Pyridine-barbituric acid colorimetry	0.002	
20	Volatile Phenol	After distillation-4-AAP spectrophotometric method	0.002	GB 7490—87
21	Petroleum	Infrared spectrophotometry	0.01	GB/T 16488—1996
22	Anionic surfactant	Methylene blue spectrophotometry	0.05	GB 7494—87
23	Sulfide	Methylene blue spectrophotometry	0.005	GB/T 16489—1996
		Direct development of the spectrophotometry	0.004	GB/T 17133—1997
24	Fecalcoliform	Naifold zymotechnics and filter membrane method		1)

Notes:

The following analysis methods will be adopted temporarily; after the national method standards are promulgated, the national standards shall be implemented

Monitoring & Analysis Method of Water and Wastewater (Version 3), China Environmental Science Press, 1989

Table 2.5 Analysis method for supplementary items of centralized domestic drinking water surface water source area

No.	Item	Analysis method	Minimal limit of detection/(mg/L)	Method origin
1	Sulfate	Weight method	10	GB 11899—89
		Flame atomic absorption spectrophotometric method	0.4	GB 13196—91
		Barium chromate spectrophotometry	8	1)
		Ion chromatography	0.09	HJ/T 84—2001
2	Chloride	Silver nitrate titrimetry	10	GB 11896—89
		Mercuric nitrate titrimetry	2.5	1)
		Ion chromatography	0.02	HJ/T 84—2001
3	Nitrate	Spectrophotometric method with phenol disulfonic acid	0.02	GB 74—87
		Ultraviolet spectrophotometry	0.08	1)
		Ion chromatography	0.08	HJ/T 84—2001
4	Iron	Flame atomic absorption spectrophotometric method	0.03	GB 11911—89
		Phenanthroline spectrophotometry	0.03	1)
5	Manganese	Flame atomic absorption spectrophotometric method	0.01	GB 11911—89
		Formaldoxime spectrophotometry	0.01	1)
		Periodate potassium spectrophotometry	0.02	GB 11906—89

Notes:
The following analysis methods will be adopted temporarily; after the national method standards are promulgated, the national standards shall be implemented
Monitoring & Analysis Method of Water and Wastewater (Version 3), China Environmental Science Press, 1989

Table 2.6 Analysis method for specific items of centralized domestic drinking water surface water source area

No.	Item	Analysis method	Minimal limit of detection / (mg/L)	Method origin
1	Trichloromethane	Headspace gas chromatography	0.0003	GB/T 17130—1997
		Gas chromatographic method	0.0006	2)
2	Carbon tetrachloride	Headspace gas chromatography	0.00005	GB/T 17130—1997
		Gas chromatographic method	0.0003	2)
3	Bromoform	Headspace gas chromatography	0.001	GB/T 17130—1997
		Gas chromatographic method	0.006	2)

(continued)

Table 2.6 (continued)

No.	Item	Analysis method	Minimal limit of detection / (mg/L)	Method origin
4	Dichloromethane	Headspace gas chromatography	0.0087	2)
5	1,2-dichloroethane	Headspace gas chromatography	0.0125	2)
6	Epichlorohydrin	Gas chromatographic method	0.02	2)
7	Chloroethylene	Gas chromatographic method	0.001	2)
8	1,1-dichloroethylene	Gas chromatography with purge and trap	0.000018	2)
9	1,2-dichloroethylene	Gas chromatography with purge and trap	0.000012	2)
10	Trichloro ethylene	Headspace gas chromatography	0.0005	GB/T 17130—1997
		Gas chromatographic method	0.003	2)
11	Tetrachloroethylene	Headspace gas chromatography	0.0002	GB/T 17130—1997
		Gas chromatographic method	0.0012	2)
12	Chloroprene	Headspace gas chromatography	0.002	2)
13	Hexachlorobutadiene	Gas chromatographic method	0.00002	2)
14	Styrene	Gas chromatographic method	0.01	2)
15	Formaldehyde	Acetylacetone spectrophotometry	0.05	GB 13197—91
		4-amino-3-hydrazino-5-mercapto-1,2,4-triazol	0.05	2)
16	Acetaldehyde	Gas chromatographic method	0.24	2)
17	Acrolein	Gas chromatographic method	0.019	2)
18	Trichloracetic aldehyde	Gas chromatographic method	0.001	2)
19	Benzene	Headspace gas chromatography	0.005	GB 11890—89
		Headspace gas chromatography	0.00042	2)
20	Methylbenzene	Headspace gas chromatography	0.005	GB 11890—89
		Carbon disulfide extraction gas chromatography	0.05	
		Gas chromatographic method	0.01	2)

(continued)

Table 2.6 (continued)

No.	Item	Analysis method	Minimal limit of detection / (mg/L)	Method origin
21	Ethylbenzene	Headspace gas chromatography	0.005	GB 11890—89
		Carbon disulfide extraction gas chromatography	0.05	
		Gas chromatographic method	0.01	2)
22	Xylene	Headspace gas chromatography	0.005	GB 11890—89
		Carbon disulfide extraction gas chromatography	0.05	
		Gas chromatographic method	0.01	2)
23	Cumene	Headspace gas chromatography	0.0032	2)
24	Chlorobenzene	Gas chromatographic method	0.01	HJ/T 74—2001
25	1,2- dichlorobenzene	Gas chromatographic method	0.002	GB/T 17131—1997
26	1,4- dichlorobenzene	Gas chromatographic method	0.005	GB/T 17131—1997
27	Trichlorobenzene	Gas chromatographic method	0.00004	2)
28	Tetrachlorobenzene	Gas chromatographic method	0.00002	2)
29	Hexachlorobenzene	Gas chromatographic method	0.00002	2)
30	Nitrobenzene	Gas chromatographic method	0.0002	GB 13194—91
31	Dinitrobenzene	Gas chromatographic method	0.2	2)
32	2,4-dinitrotoluene	Gas chromatographic method	0.0003	GB 13194—91
33	2,4,6-trinitrotoluene	Gas chromatographic method	0.1	2)
34	Nitrochlorobenzene	Gas chromatographic method	0.0002	GB 13194—91
35	2,4-dinitrochlorobenzene	Gas chromatographic method	0.1	2)
36	2,4-dichlorophenol	Electron capture-capillary chromatography	0.0004	2)
37	2,4,6-trichlorophenol	Electron capture-capillary chromatography	0.00004	2)

(continued)

Table 2.6 (continued)

No.	Item	Analysis method	Minimal limit of detection / (mg/L)	Method origin
38	Pentachlorophenol	Gas chromatographic method	0.00004	GB 8972—88
		Electron capture-capillary chromatography	0.000024	2)
39	Phenylamine	Gas chromatographic method	0.002	2)
40	Benzidine	Gas chromatographic method	0.0002	3)
41	Acrylamide	Gas chromatographic method	0.00015	2)
42	Acrylonitrile	Gas chromatographic method	0.10	2)
43	Dibutyl phthalate	Liquid chromatography	0.0001	HJ/T 72—2001
44	Bis (2-ethylhexyl) phthalate	Gas chromatographic method	0.0004	2)
45	Hydrazine hydrate	Dime-tylaminobenzaldchyde spectrophotometric method	0.005	2)
46	Lead tetraethyl	Dithizone colorimetry	0.0001	2)
47	Pyridine	Gas chromatographic method	0.031	GB/T 14672—93
		Barbituric acid spectrophotometric method	0.05	2)
48	Turpentine	Gas chromatographic method	0.02	2)
49	Picric acid	Gas chromatographic method	0.001	2)
50	Butyl xanthic acid	Copper reagent-Cuprous spectrophotometric method	0.002	2)
51	Active chlorine	N,N-Diethyl-1,4-phenylenediamine spectrophotometry	0.01	2)
		3,3′,5,5′- tetramethyl benzidine colorimetric method	0.005	2)
52	DDT (Dichlorodiphenyl)	Gas chromatographic method	0.0002	GB 7492—87
53	Lindane	Gas chromatographic method	4×10^{-6}	GB 7492—87
54	Heptachlor epoxide	Liquid-liquid extraction gas chromatographic method	0.000083	2)

(continued)

Table 2.6 (continued)

No.	Item	Analysis method	Minimal limit of detection / (mg/L)	Method origin
55	Parathion	Gas chromatographic method	0.00054	GB 13192—91
56	Parathion-methyl	Gas chromatographic method	0.00042	GB 13192—91
57	Malathion	Gas chromatographic method	0.00064	GB 13192—91
58	Dimethoate	Gas chromatographic method	0.00057	GB 13192—91
59	DDVP	Gas chromatographic method	0.00006	GB 13192—91
60	Dipterex	Gas chromatographic method	0.000051	GB 13192—91
61	Demeton	Gas chromatographic method	0.0025	2)
62	Chlorothalonil	Gas chromatographic method	0.0004	2)
63	Carbaryl	High performance liquid chromatography	0.01	2)
64	Deltamethrin	Gas chromatographic method	0.0002	2)
		High performance liquid chromatography	0.002	2)
65	Atrazine	Gas chromatographic method		3)
66	Benzopyrene	Acetylated paper chromatography fluorescence spectrophotometric method	4×10^{-6}	GB 11895—89
		High performance liquid chromatography	1×10^{-6}	GB 3198—91
67	Methyl mercury	Gas chromatographic method	1×10^{-8}	GB/T17132—1997
68	Polychlorinated biphenyl	Gas chromatographic method		3)
69	Microcystic toxins—LR	High performance liquid chromatography	0.00001	2)
70	Yellow phosphorus	Mo-Sb Anti Spectrophotometric Method	0.0025	2)
71	Molybdenum	Flameless atomic absorption spectrophotometry	0.00231	2)
72	Cobalt	Flameless atomic absorption spectrophotometry	0.00191	2)

(continued)

Table 2.6 (continued)

No.	Item	Analysis method	Minimal limit of detection / (mg/L)	Method origin
73	Beryllium	Spectrophotometric method with eriochrome cyanine R	0.0002	HJ/T 58—2000
		Graphite furnace atomic absorption spectrophotometry	0.00002	HJ/T 59—2000
		Morin fluorescence spectrophotometric method	0.0002	2)
74	Boron	Curcumin spectrophotometric method	0.02	HJ/T 49—1999
		Methylenimine-H spectrophotometric method	0.2	2)
75	Antimony	Atomic absorption spectrophotometry	0.00025	2)
76	Nickel	Flameless atomic absorption spectrophotometry	0.00248	2)
77	Barium	Flameless atomic absorption spectrophotometry	0.00618	2)
78	Vanadium	Tantalum reagent (BPHA) extraction spectrophotometry	0.018	GB/T 15503—1995
		Flameless atomic absorption spectrophotometry	0.00698	2)
79	Titanium	Catalytic oscillopolarographic method	0.0004	2)
		Salicyl fluorine spectrophotometry	0.02	2)
80	Thallium	Flameless atomic absorption spectrophotometry	1×10^{-6}	2)

Notes:
The following analysis methods will be adopted temporarily; after the national method standards are promulgated, the national standards shall be implemented
(1) Monitoring & Analysis Method of Water and Wastewater (Version 3), China Environmental Science Press, 1989
(2) Specification for Hygiene of Drinking Water, Ministry of Health of the People's Republic of China, 2001
(3) Standard Inspection Method of Water and Wastewater (Version 15), China Building Industry Press, 1985

7. Standard Implementation and Supervision

7.1 The standard shall be implemented and supervised by the administrative competent department of environment protection and related department of local people's government at or above county level in accordance with duties, responsibilities and division of labor.

7.2 The standard-exceeding indexes of the water quality of the surface water resource places of concentrated drinking water must meet the Requirements of the *Hygienic Specifications for Drinking Water* after they are purified in running water plants.

7.3 For the indexes that are not stipulated in the standard, the people's governments of provinces, autonomous regions and metropolitans under central government can formulate local standards and reports to the administrative competent department of the environment protection under State Council for record.

Attachment 2.2. The Implementation Guidelines of the Protective Regulations of the Water Source of the Upper Reaches of Huangpu River in Shanghai

(*The Implementation Guidelines of the Protective Regulations of the Water Source of the Upper Reaches of Huangpu River in Shanghai* was issued by Shanghai people's government on August 29, 1987, revised for the first time pursuant to the *Division of Shanghai People's Government on the Revision of the Implementation Guidelines of the Protective Regulations of the Water Source of the Upper Reaches of Huangpu River in Shanghai* on May 28, 1996, and revised for the second time and reissued pursuant to the order No. 53 of Shanghai people's government.)

Chapter One General Rules

Article 1 This Guidelines is formulated in accordance with the *Protective Regulations of the Water Source of the Upper Reaches of Huangpu River in Shanghai* (hereafter referred to as regulations).

Article 2 The specific scope of the water resources protection zone specified in the second article of regulations is:

The upper boundary of north bank is the boundary between Dianshan Lake—Shanghai, Jiangsu; the lower boundary of north bank is Minhang west boundary—Xihebang; the trend line of the five kilometers (in depth land) scope boundary of north bank: Wujiaxiang—Hejiatang— Lujiabang—Sanzhuang (Wangjiatang)—Zhejialou—Hejiatang—(Huayang)—Dongmen (Caijia Village)—Zhongshan Road 1 of Songjiang County—Zhoujia Village—Yaojiabang—Zhao- zhuang—Datiebang—Wangjin(Wangjiabang)—Wangxin—Nima (Nijiabang)—Xiaogang—Ding- jiabang—Henggang—Beidangjing.

The upper boundary of south bank is the boundary between Dianshan Lake—Shanghai, Jiangsu; the lower boundary of south bank is Qianbujing; the trend line of the five kilometers (in depth) land scope boundary of south bank: Tangxifang—Jinjiadai—Cheting—Xiaojiadai— Maqiao—Shanfang—Yangjiayu—Wenhe—Huangnibang—Shengli(Dongyaotan)—Yanghebang—Rutang(Qijiadai)—Nanjiejing—Yangjiadian—Sihe—Zhengdong (Xixiaozhen)—Nanhenggang— Xiyieku—Gangdu—Jiagang.

Damao Port, Yuanxiejing, Taipu River upstream ten kilometers waters scope is:

(1) From the confluence of Hengliaojing, Shuliaojing and Damao Port to Dujiabang of Jueshi Port;
(2) From the confluence of Hengliaojing, Shuliaojing and Damao Port to Qianjin of Xiaomao Port;
(3) From the confluence of Xietang, Hengliaojing and Yuanxiejing to Yangli of Yuanxiejing;
(4) From the confluence of Taipu River and Ximao River to the boundary of Shanghai.

Article 3 The specific scope of the quasi water resources protection zone specified in the Second Article of Regulations is:

The upper boundary of north bank and west bank is Minhang west boundary—Xihebang; the lower boundary of north bank and west bank is Longhua Port—Caobao Road; the trend line of the five kilometers (in depth) land scope boundary of north bank and west bank is: Chuanbang—Huaer (Caojiatang)—Longxi (Zhangjia Xinzhai)—Hangxi—(Mujiayan)—Guang- hui—Xiangyang—Guangming (Zhoujiazhai)—Xiangzhali—Beiqiao Town—Wujia Port.

The upper boundary of south bank and east bank is Qianbujing; the lower boundary of south bank and east bank is Chuanyang River; the trend line of the five kilometers (in depth) land scope boundary of south bank and east bank is: Zhujiazhai—Yangjiazhai—Zhang- jiabang—Nanbu(Dongtangjiazhai)—Tianhua'an—Zhixin (Xujiali)—TiqiaoTown—Sunjiazhai—Jiashanzhai—Dongtaizhai—Huibei(Yejiazhai)—Huihong—Guangji—Jiguang—Yejiaku—Zhu- jiatang—Zhejiatang—Dingjia—Chenjiawan—Wujiazhai—Houjiazhai—Chenghebang—Malu—Wangjiali—Tangxifang.

Article 4 Where there are roads, highways, rivers within the five kilometers (in depth) land (±200 meters) along two banks of the rivers and lakes of water resources protection areas and quasi water resources protection areas, the boundary is the side of roads, highways and rivers far from Huangpu River and Dianshan Lake.

The specific boundary of water resources protection areas and quasi water resources protection areas is determined by Shanghai Urban Planning Administration (hereafter referred to as Planning Administration) and Shanghai Environment Protection Bureau (hereafter referred to as Environment Protection Bureau).

Article 5 "The water from upper reaches" in Article 3 of Specifications means Jishiu Port, Taipu River, Yuanxiejing and other tributaries that run into water resources protection areas.

The water quality of each tributary that runs into quasi water resources protection areas must not be lower than national environment quality standard of Class III surface water in order to ensure it to reach the national environment quality standard of Class III surface water.

Article 6 The main duties and responsibilities of Environment Protection Bureau are:

(1) To organize and exercise Regulations, improve the specific protective measures of Shanghai for the water resources of the upper reaches of Huangpu River, uniformly manage the protection of the water resources of the upper reaches of Huangpu River;
(2) To organize, coordinate, supervise and implement water pollution prevention and control and plan of the water resources protection areas and quasi water resources protection areas;
(3) To do well the work of the joint prevent and control of the water pollution of Dianshan Lake, Yuandang Lake, Jishiu Port, Taipu River, Yuanxiejing, Damao Port, etc., in cooperation with Tai Lake Reaches Administration and negotiation with Jiangsu and Zhejiang provinces and make joint prevention and control plan for water pollution and enact implementation measures;
(4) To establish the water environment monitoring network of the water resources protection areas and quasi water resources protection areas for water quality monitoring to collect monitoring data and understand water quality development;
(5) To take charge to the implementation of emission permit system and supervise limited period control of pollution sources;
(6) To be responsible for the review and approval of the simultaneous design, simultaneous construction and simultaneous put into production (here after referred to as three simultaneousnesses) of the facilities pollution prevention and control and other public nuisances and the main parts of projects;
(7) To sum up and spread the advanced experiences and technologies of water pollution prevention and control and reward the advanced in environment protection;
(8) To investigate and handle water pollution accidents;
(9) To collect emission cost, standard-exceeding emission cost.

Article 7 Supervision institutions at different levels are responsible for the supervision and control of ship emission. Their main duties and responsibilities are:

(1) To carry out and perfect the specific measures of ship pollution prevention and control within the water environment monitoring network of the water resources protection area and quasi water resources protection area of the upper reaches of Huangpu River;
(2) To supervise ship pollution, investigate and handle water pollution accidents caused by ships;
(3) To monitor land water pollution in cooperation with environment protection department;

(4) To sum up and spread the advanced experiences and technologies of ship pollution prevention and control and reward the advanced units and individuals in environment protection;
(5) To periodically report the implementation of Regulations to Environment Protection Bureau.

Article 8 The administrative Departments of planning, water conservation, municipal administration, health and public utilities, etc., shall exercise Regulations in cooperate with environment protection departments. Their main duties and responsibilities are as follow:

The planning department, in the place choice for county plans, regional plans, town plans and construction projects, where the water resources protection area of the upper reaches of Huangpu River is involved, must take the requirements of water resources protection into consideration; reasonably adjust the industrial pattern and structure of the upper reaches of Huangpu River in cooperation with environment protection departments and industrial competent departments; implement three simultaneousnesses review in cooperation with environment protection departments.

The water conservation department shall protect the water resources quality of the upper reaches of Huangpu River in cooperation with environment protection departments; when adjusting and allocating water resources, must guarantee certain flow to protect water environment quality; provide the hydrological data of the upper reaches areas of Huangpu River to environment protection departments.

Municipal administration shall perfect the drainage and sewage treatment system of the upper reaches areas of Huangpu River to improve sewage treatment rate and pollutants- removing rate; formulate urban facilities management measures, investigate and handle the sewage emission accidents damaging drainage facilities.

The department of fishery administration shall enact and perform the management measures to prevent fishery production from polluting the water resources of the upper reaches of Huangpu River; take part in investigating and handling the water pollution accidents caused by fishery production; supervise and manage fishery ship emission in cooperation with port and navigation supervision institutions.

Health department shall participate the three simultaneousnesses review of the sanitation protection facilities related to new projects, extension projects and reconstruction projects of the upper reaches areas of Huangpu River as well as the investigation of related serious pollution accidents.

Public utilities department shall carry out regular monitoring of the water quality of the water coming in and going out of water plants and provide monitoring data to environment protection departments.

Article 9 People's governments at all levels shall urge the competent departments of enterprises, institutions, planning and environment protection departments, etc., to strictly exercise regulations. Effective measures shall be taken to reduce the emission amount of the existing water pollution in the water resources protection areas and quasi water resources protection area. New water pollution shall be strictly prevented.

Chapter Two Supervision and Management

Article 10 Environment Protection Bureau shall propose annual plan for total pollutant emission reduction in accordance with environment protection plan, the requirements of the water quality of the water resources protection areas and quasi water resources protection areas and regional control indexes of total pollutant emission, and targeted at the different characteristics of various areas and industries.

Article 11 A pollutants-emitting unit shall implement the corresponding annual plan for total pollutant emission pursuant to its control indexes of total pollutant emission determined by environment protection department.

Article 12 Each district and county can moderately develop some projects with little pollution in the areas where actual total pollutant emission is lower than allowed total pollutant emission and where water quality reaches water environment quality standard specified in water resources protection, but the production projects with serious pollution, such as asbestos products, soil sulfur, electrolytic plating, leather making, paper making, pulp making, soil coking, bleaching and dyeing, oil refining, nonferrous metal metallurgy, soil phosphate fertilizer, dye and the like must not be constructed; oil projects, wharves for hazardous products and scraping yards must not be built. The emission of new projects must be controlled within local allowed total pollutant emission indexes.

Total pollutant emission indexes can be comprehensively balanced within areas, adjusted for more or less between enterprises, transferred mutually. But this must be approved by environment protection departments.

Article 13 A pollutants-discharging unit must apply for pollutant emission to the environment protection department of the district or county where it is. The emission application shall include pollutant types, emission concentration, quantity, time, emission outlet location, pollution control measures and the period in which allowed emission is reached. Pollutants are not allowed to be discharged without permission.

The environment protection department shall inspect the emission application of each unit in accordance with the control requirements of total pollutant emission, award Shanghai Enterprise and Institution Permit for Water Pollutant Emission to whatever unit that reaches the control requirements of total pollutant emission and Shanghai Enterprise and Institution Permit for Temporary Water Pollutant Emission to whatever unit that does not reach the control requirements of total pollutant emission and order it to control within limited period.

Article 14 Sewage metering instruments must be installed on the sewage emission outlets within the water resources protection areas and quasi water resources protection areas. The sewage discharging unit shall have environment protection managements to guarantee the normal operation of sewage treating facilities and take original record. Regular reports shall be submitted to environment protection departments; cheating is not allowed.

For the sewage treating facilities that can not normally operate because of faults and overhaul, etc., the sewage discharging unit shall take emission reduction measures, such stopping or reducing production, etc., and report them to the environment protection department of the district or county where it is. Where it is necessary to scrap or stop the use of the sewage treating facilities, the unit shall make an application in writing before scraping or stopping use to the environment protection department.

Article 15 The projects that causes water pollution must not be built within five kilometers (in depth) land area along Dianshan Lake, Yuandang Lake.

New projects, extension projects and reconstruction projects located within 2–5 kilometers (in depth) land area along Dianshan Lake, Yuandang Lake and within the land areas of the other water resources protection areas and quasi water resources protection areas that discharge pollutants into water bodies shall meet specified emission standards and determined control indexes of total pollutant emission (specific emission standards attached below).

Sanatoriums and scenic sports can be built within 0.2–2 kilometers (in depth) land area along Dianshan Lake, Yuandang Lake and within the land areas, but must conform to the overall development plan of Dianshan Lake and Yuandang Lake, equipped with corresponding pollution control facilities, and discharge the domestic sewage that reaches emission standards to the appointed waters outside two kilometers (in depth) along lakes. Where it is necessary to discharge sewage on site, they have to reach national environment quality standard of Class II surface water.

In addition to open greening belts, any other project must not be built within 0.2 kilometer (in depth) land area along Dianshan Lake, Yuandang Lake.

Article 16 Emission outlets must not be established along lake shores of Dianshan Lake and Yuandang Lake.

Article 17 In addition to normal navigational channel maintenance projects and water conservation projects, any new construction project and on-water activities project must not be developed within Dianshan Lake and Yuandang Lake. Ships that are not equipped with pollutants storing devices and containers in accordance with provisions are not allowed to sail and carry out tasks on Dianshan Lake and Yuandang Lake.

Chapter Three Water Resources Pollution Prevention and Control

Article 18 It is banned to pile, dump and bury coal ash, slag, tailings, foots, radioactive substances, poisonous and harmful articles and other industrial, building and life solid wastes within the water resources protection areas and quasi water resources protection areas; those that have been piled must be cleaned by the environment protection departments of districts and counties within limit period.

Temporarily piling slags, garbage and other non-poisonous and harmless solid wastes must adopt strict protective measures, must be applied to the environment protection departments of districts and counties and approved by them. Temporarily piling time is not more than half year at most; they must be cleaned upon expiry.

Article 19 Highly toxic and high-persistent pesticide, such as benzene hexachloride, DDT, 1605, 1059, organic mercury compounds, etc., may not be used within the water resources protection areas and quasi water resources protection areas of the upper reaches of Huangpu River.

Article 20 For the storage and transport of fertilizer and pesticide, related safety management provisions must be exercised. The fertilizer and pesticide that are ineffective and banned must not be dumped into the water bodies of Hangpu River water system. The water bodies of Hangpu River water system must not be used to clean the packaging materials, containers and carrying tools.

Article 21 The following ship tasks must be applied to and approved by port and navigation supervision institutions in advance.

(1) Ship cabins clean up;
(2) Discharging cabins-ballasting water, cabins-cleaning water and ship bottom water;
(3) Oily tasks;
(4) Flushing the decks and cabins of the ships carrying poisonous, harmful or dust-raising bulk cargoes;
(5) Painting ship shells;
(6) Handling oily waste water.

Article 22 All ships must be equipped with corresponding pollution preventing equipments and instruments. Ships-building units must have necessary pollution preventing equipments and instruments, adopt measures in the course of tasks to strictly prevent oil, oily mixtures and wastes from falling into rivers.

Article 23 New projects and extension projects that lead to water pollution must not be constructed within the waters of two kilometer river section respectively of the upper and lower reaches at the two sides of new running water collecting outlets. In addition to hydrological and water quality investigating and monitoring ships, other ships are strictly banned to carry out loading and unloading tasks. Existing sewage discharging outlets shall be controlled or moved within limited period.

Article 24 Reclaiming lakeland is strictly banned within the water resources protection areas; mechanically catching shellfish and hunting wild aquatic birds and other activities damaging the ecological balance of water environment are banned; destructing trees and vegetation and killing beneficial birds, animals and insects, etc., are not allowed.

The existing aquafarms in lakes within the water resources protection areas must be managed more vigorously and may not be expanded without authorization.

Fertilizers that pollute water, such as human feces and bovine dung, etc., are banned to be used in net cage culture areas.

New aquafarms are banned to be built in Dianshan Lake and Yuandang Lake.

Chapter Four Award and Punishment

Article 25 A pollution-discharging unit can apply the reduction and exemption of product tax or value-added tax to tax authorities in accordance with the provisions of Ministry of Finance's Notice for the Related Taxation Problem of Exceeding Comprehensively Using Resources ((1985) Finance and Tax Document No. 334) for the products produced by the projects of comprehensively using resources that are constructed by it with its own capital; it can be exempt from income tax and regulation tax within five years for the newly-added profit of its workshops and branches with independent accounting and responsible for their own profit and loss after they are put into production. The tax-free part of profit is given to the enterprise for special fund of environment protection.

Article 26 The units and individuals that perform one of the following actions will be praised and rewarded by environment protection departments or port and navigation supervision institutions.

(1) To finish waste water prevention and control projects in advance with notable environment effect.
(2) To inform and disclose water pollution accidents and the behaviors against Regulations and the Guidelines and be real through verification.
(3) To take effective measure to avoid water pollution accidents or lessen the damage of water pollution accidents.
(4) To propose reasonable proposals for water pollution prevention and control, or have technical innovations and inventions that obviously reduce emission after they are adopted.
(5) The total pollutant emission discharged by an enterprise is lower than allocated control indexes, which is real after verification.
(6) To restore, strengthen the ecological balance of upper reaches to improve regional environment quality with obvious effect.

Article 27 Whatever unit or individual that meet the conditions specified in Article 26 of the Guidelines will be reported by the environment protection departments of districts and counties or competent authorities to Environment Protection Bureau for praise and award. The award will come from standard-exceeding emission cost.

Article 28 The units and individuals that perform one of the following actions will be given the following punishments by environment protection departments or port and navigation supervision institutions in accordance with the following provisions.

(1) Not to make the application for pollutant emission exceeding time limit, warns or fine not more than RMB 300 yuan.
(2) To have excessive emission against the requirements specified in permit for emission, two or three times fine, not less than RMB 1,000 yuan at least but not more than RMB 200 thousand yuan at most.
(3) To construct new projects, extension project and reconstruction project that break the Article 8 of Regulations and for which the provisions of the environment protection of the construction projects are not carried out, below RMB 5,000 yuan if the emission does not exceed the specified control indexes of total pollutant emission of the unit itself, RMB 5,000–10,000 yuan if the emission exceeds the specified control indexes of total pollutant emission of the unit itself.
(4) To refuse and hinder the on-site inspection of environment protection departments or tell lies, warns or fine not more than RMB 5,000 yuan.
(5) To control under limited time period but not meet the specified requirements exceeding the time limit, fine of two-three times standard-exceeding emission cost, not less than RMB 1000 yuan at least but not more than RMB 200 thousand yuan at most.
(6) To break the provisions of Article 12 (II), Article 13 and 14 of Regulations, fine not more than RMB 20,000 yuan.
(7) To Violate the provisions of Article 12 (I), Article 15 (I), Article 17 (I) or (II), Article 18 of Regulations, fine not more than RMB 50,000 yuan.
(8) To violate the provisions of other articles of Regulations, fine not more than RMB 10,000 yuan

Article 29 Units, the people in charge of ships and the people in charge of direct responsibility will be fined not more than RMB 200 yuan by environment protection departments or port and navigation supervision institutions if the units break Regulations and the Guidelines.

Article 30 The workers of environment protection departments or port and navigation supervision institutions who misuse their authority, break the law in the knowledge of the law, make use of their position as officials to serve private ends, practise fraud, so that the implementation of Regulations is affected, will be given administrative punishment pursuant to the provisions of the awards and punishments of the workers of national administrative organs.

Article 31 The fine of enterprise units shall be paid from retained profit after tax or enterprise funds; the fine of institution units shall be paid from the balance of budget.

Chapter Five Supplementary Provisions

Article 32 The Guidelines is interpreted by Environment Protection Bureau in the specific application of it.

Article 33 The Guidelines was put into effect on September 1, 1987.

Appendix 3

Attached Figures 1. Overall Plan Maps of Shanghai Dalian Lake Wetland Restoration Project (2,000 mu)

Fig. 1.1 Current land use map of the project

	Functional zoning	Area (hm^2)
A	Wetland restoration and reconstruction area	30
B	Forest wetland cultivation area	25
C	Shoal wetland restoration and biodiversity conservation area	10
D	Wetland organic agricultural area	45
E	Coastal wetland restoration area	5
F	Community domestic sewage treatment area	15

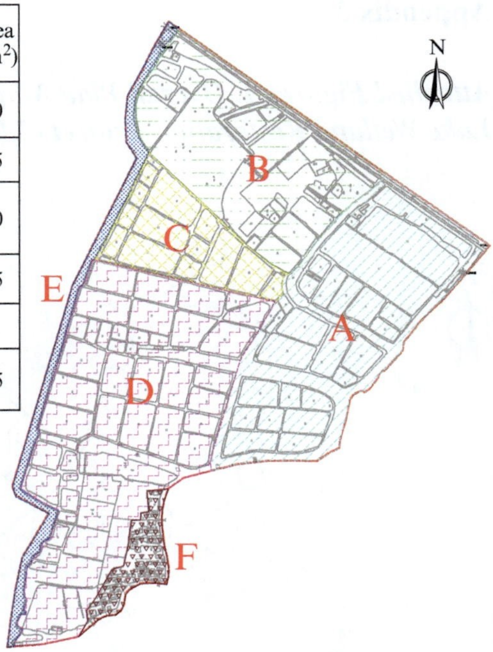

Fig. 1.2 Functional zoning map of the project

Fig. 1.3 Current drainage map

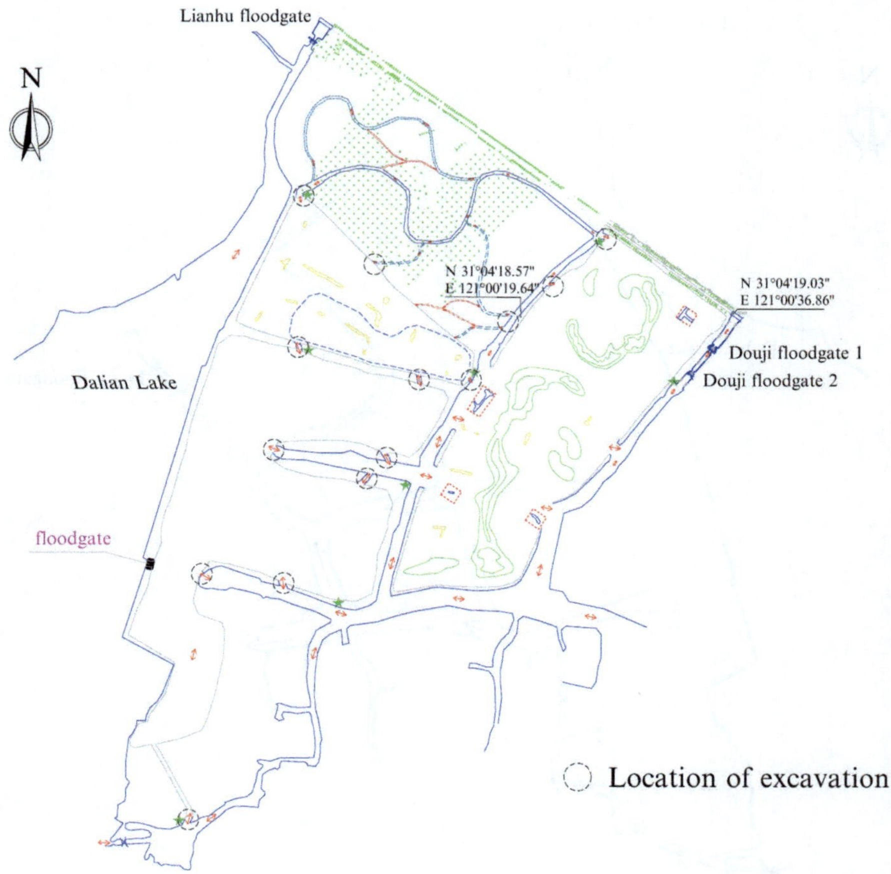

Fig. 1.4 Design drainage map

Fig. 1.5 Infrastructure map

Landtype	Area/m²	Altitude/m	Sludge deep/cm
T1	14393	0.71	18
T2	33547	1.12	24
T3	1079	1.42	23
T4	42040	1.05	16
T5	2969	0.43	17
T6	2910	0.46	19
T7	6864	1.14	21
T8	6828	1.15	23
T9	8722	1.24	20
T10	4758	1.41	26
T11	3614	1.11	19
T12	9460	1.58	16
T13	3010	1.46	18
T14	11692	0.93	20
T15	11145	1.43	17
T16	13455	1.45	27
T17	1556	0.95	31
T18	1938	1.02	20
T19	1594	1.12	21
T20	9324	1.47	24
T21	4488	1.35	26
T22	5094	1.40	21
T23	9702	1.45	19
T24	5242	1.46	15
T25	5155	1.25	14
T26	10185	1.49	13
T27	6555	1.65	19
Subtotal	237319		
Other	135081		
Total	372400		

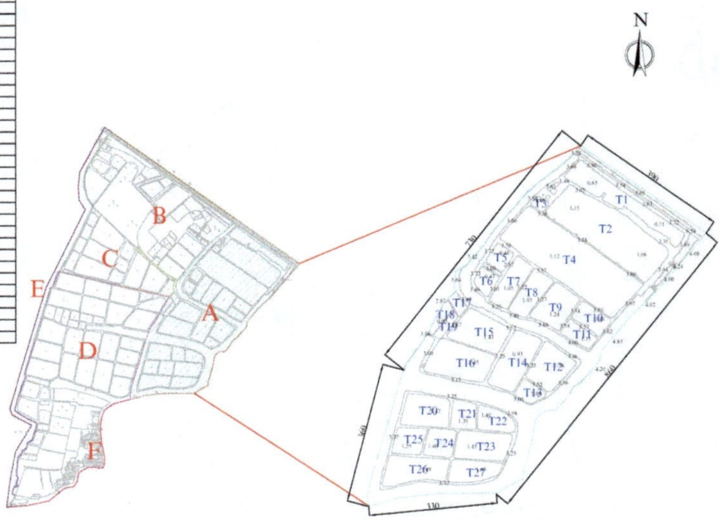

Fig. 1.6 Current map of A area

Fig. 1.7 Landform rebuilding map of A area

Fig. 1.8 Plant configuration map of A area

Landtype	Area/m²	Altitude/m	Sludge deep/cm
T28	1719	1.57	20
T29	921	1.46	23
T30	1869	1.34	25
T31	770	1.11	21
T32	8035	1.14	19
T33	9677	0.99	25
T34	9078	0.78	18
T35	1168	1.81	17
T36	2755	1.52	24
T37	702	1.72	21
T38	6211	1.76	23
T39	2250	1.87	18
T40	4232	1.93	19
T41	1877	1.92	18
T42	3872	1.11	17
T43	2887	1.02	20
T44	13988	0.31	23
T45	4233	0.80	24
T46	2487	1.41	22
Subtotal	78731		
L1	9515	2.86	
L2	62937	3.14	
L3	43857	3.37	
L4	22725	3.24	
Subtotal	139034		
Other	73235		
Total	291000		

Fig. 1.9 Current map of B area

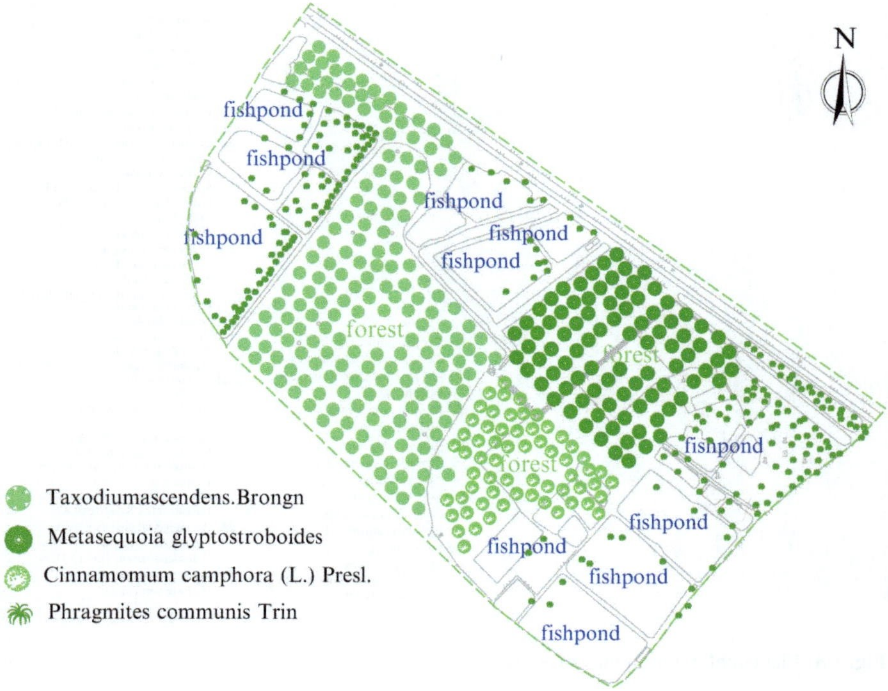

Fig. 1.10 Current plant map of B area

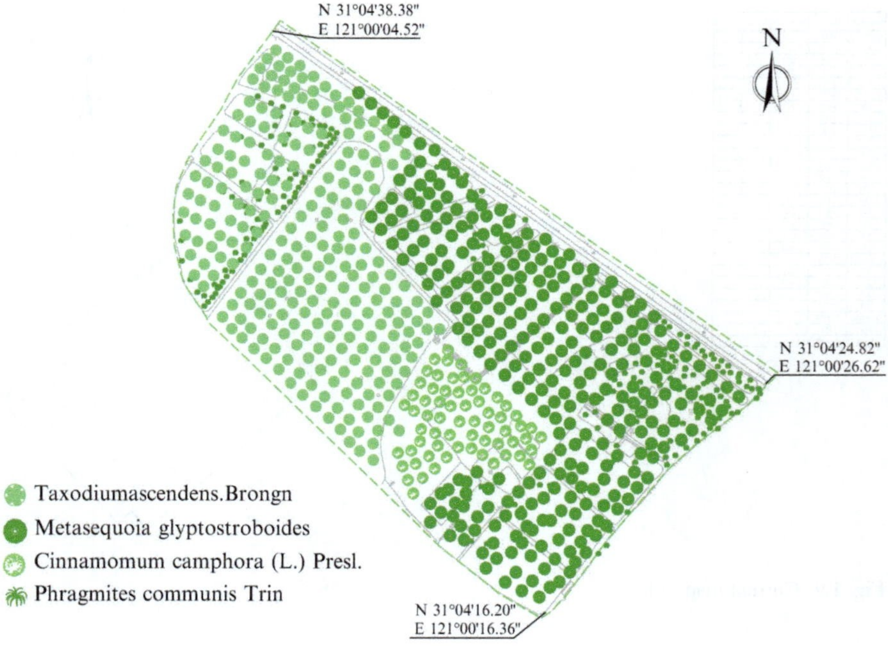

Fig. 1.11 Plant configuration map of B area

Landtype	Area/m²	Altitude/m	Sludge deep/cm
T47	3874	1.34	21
T48	12719	1.12	23
T49	13921	1.08	19
T50	4142	1.18	19
T51	3973	1.23	20
T52	4886	1.03	16
T53	17148	1.12	15
T54	13431	1.22	19
T55	14873	1.27	22
T56	561	1.10	24
T57	4241	1.10	21
T58	3278	1.07	14
T59	2855	0.97	12
T60	2620	0.74	13
T61	943	1.13	11
Subtotal	103465		
Other	20535		
Total	124000		

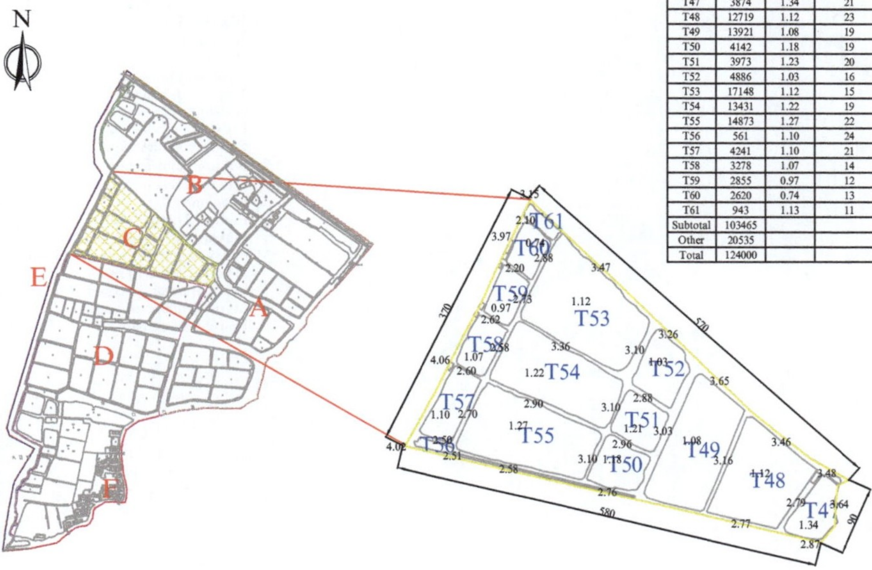

Fig. 1.12 Current map of C area

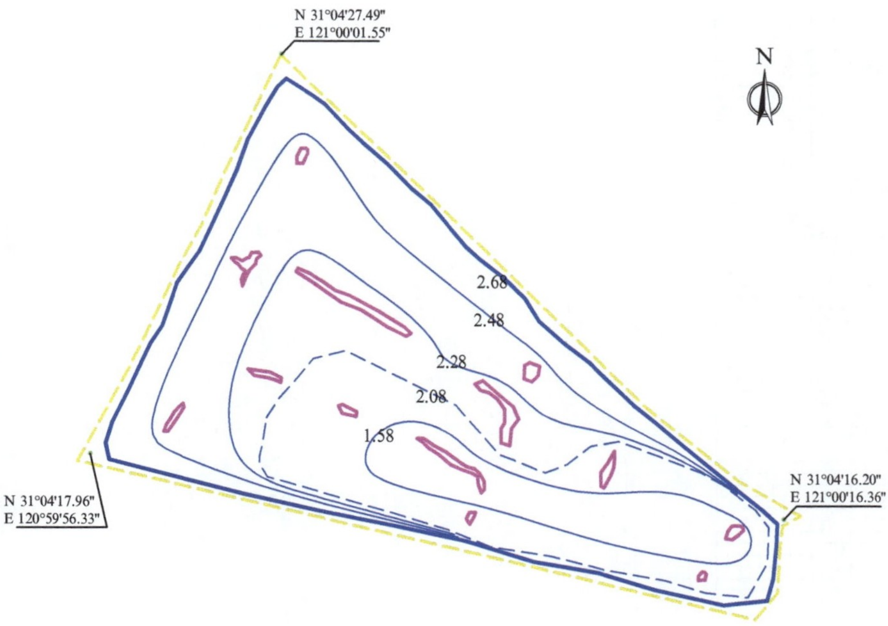

Fig. 1.13 Landform rebuilding map of C area

Fig. 1.14 Plant configuration map of C area

Landtype	Area/m²	Altitude/m	Sludge deep/cm
T62	11654	1.00	15
T63	10913	1.04	16
T64	11677	1.04	19
T65	2592	1.74	20
T66	7103	1.49	22
T67	12571	1.37	21
T68	11386	1.07	18
T69	11484	1.07	15
T70	13948	1.07	13
T71	3420	1.44	11
T72	2523	1.42	20
T73	2739	1.43	25
T74	2061	1.43	28
T75	1774	1.44	21
T76	2712	1.35	26
T77	1141	1.43	19
T78	1919	1.34	18
T79	13844	1.27	19
T80	10537	1.17	17
T81	11053	1.09	29
T82	11892	1.24	28
T83	6880	1.13	20
T84	7732	1.12	20
T85	6281	1.07	25
T86	9539	1.24	28
T87	13352	1.39	27
T88	12820	1.27	29
T89	10433	1.45	32
T90	12993	1.16	12
T91	13691	1.34	10
T92	2747	1.12	14
T93	6246	1.07	15
T94	2634	1.56	11
T94	996	1.57	16
T96	8521	1.63	17
T97	3586	1.65	13
T98	1697	1.74	15
T99	2901	1.73	14
T100	2749	1.54	13
T101	1983	1.80	15
Subtotal	286724		
D1	38551	3.62	
D2	46473	3.65	
D3	23586	3.90	
Subtotal	108610		
Town area	32892		
Other	106174		
Total	534400		

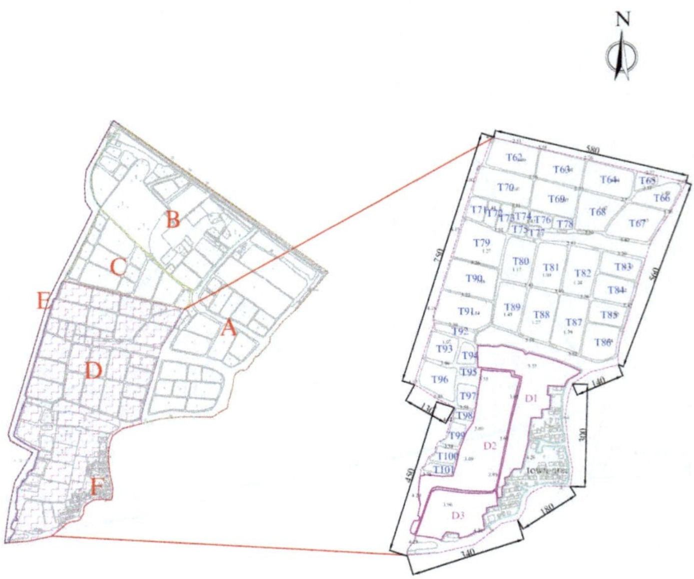

Fig. 1.15 Current map of D, F areas

Fig. 1.16 Plant configuration map of D area

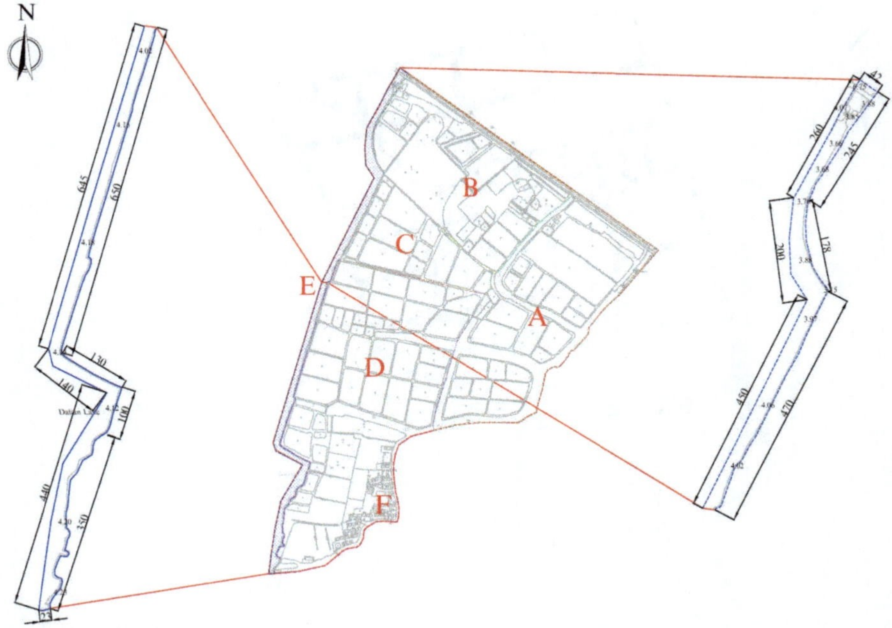

Fig. 1.17 Current map of E area

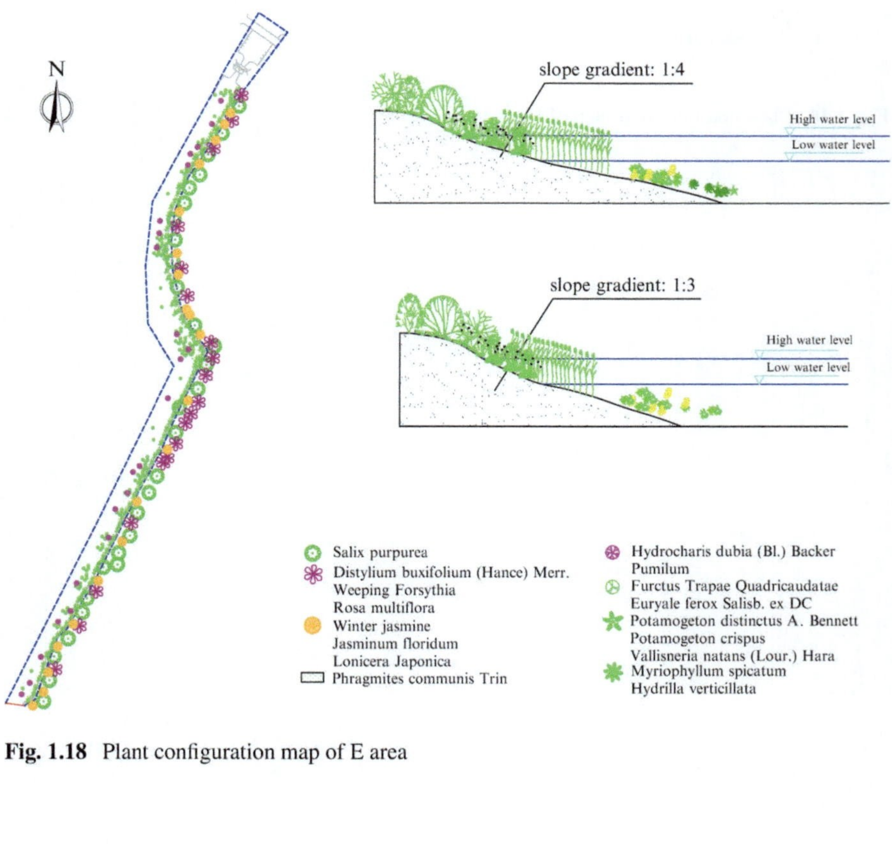

Fig. 1.18 Plant configuration map of E area

Attached Figures 2. Design Drawings of First-Stage Wetland Restoration Project (625 mu)

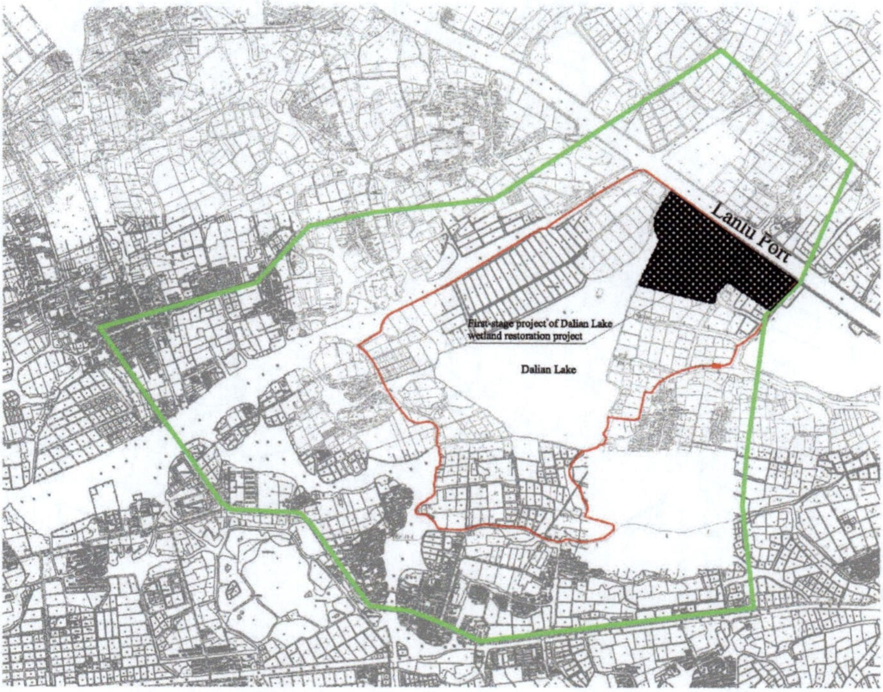

Fig. 2.1 Location of the project

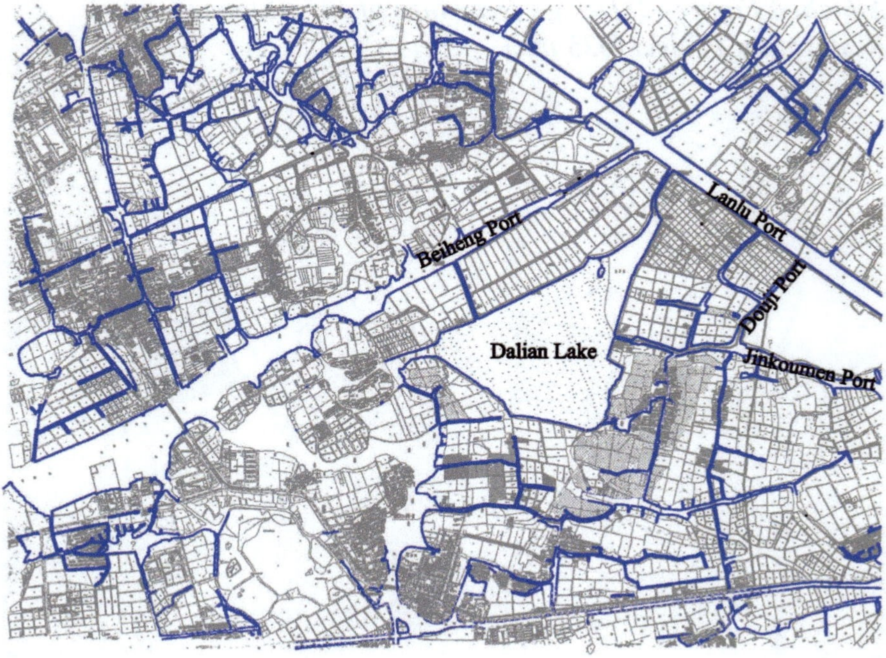

Fig. 2.2 Regional drainage map

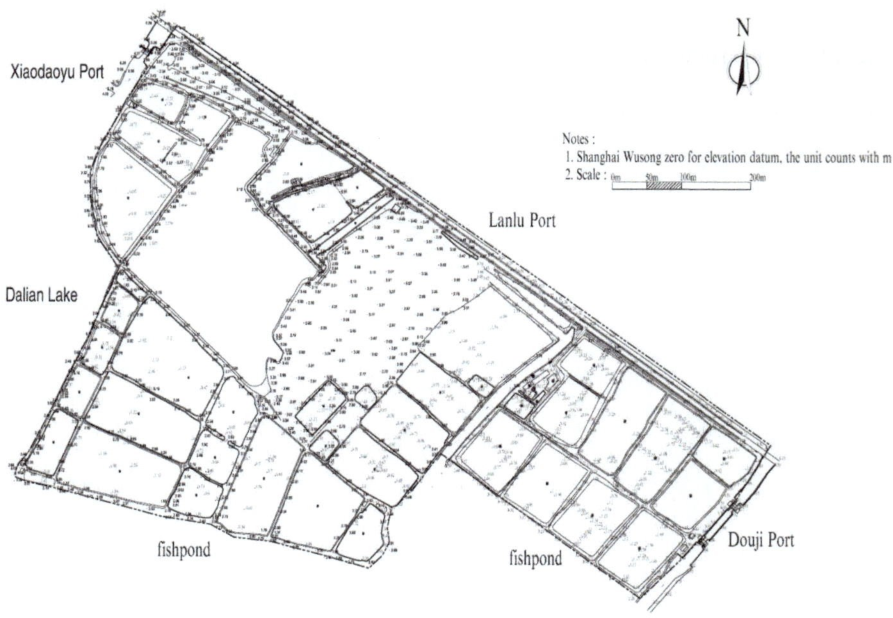

Fig. 2.3 Current landform map: *bottom* elevation of sludge

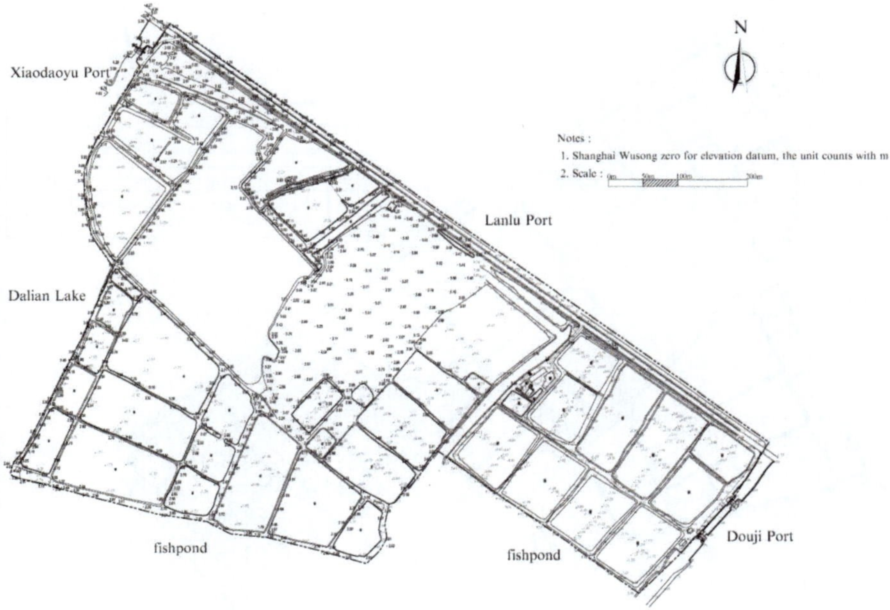

Fig. 2.4 Current landform map: *top* elevation of sludge

Landtype	Area/m^2	Percentage/%
fishpond	220277	52.9
forest	97978	23.5
bank	67300	16.2
road	6516	1.6
river	4685	1.1
other	19910	4.7
total	416666	100

Fig. 2.5 Current land use map

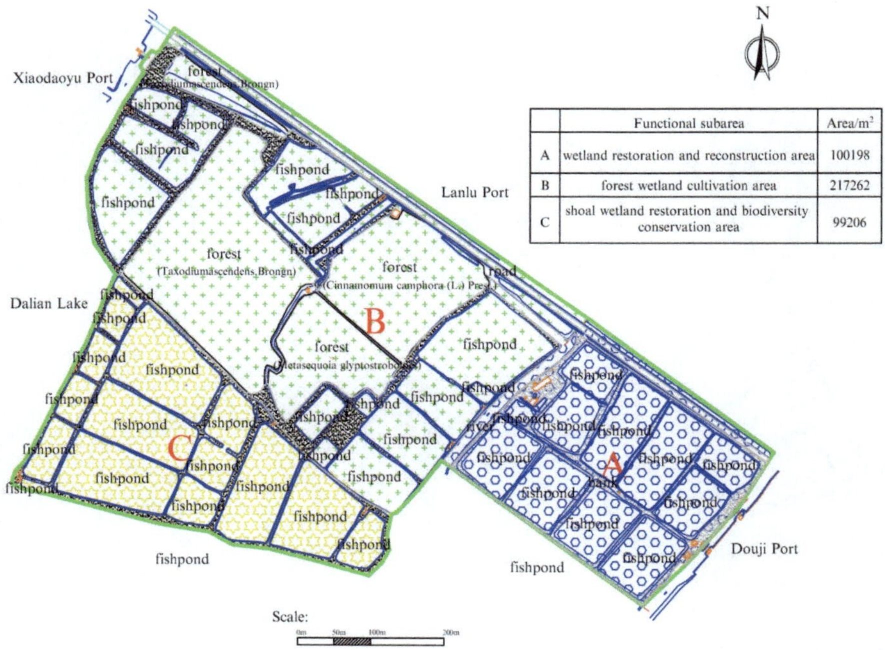

	Functional subarea	Area/m^2
A	wetland restoration and reconstruction area	100198
B	forest wetland cultivation area	217262
C	shoal wetland restoration and biodiversity conservation area	99206

Fig. 2.6 Functional zoning map

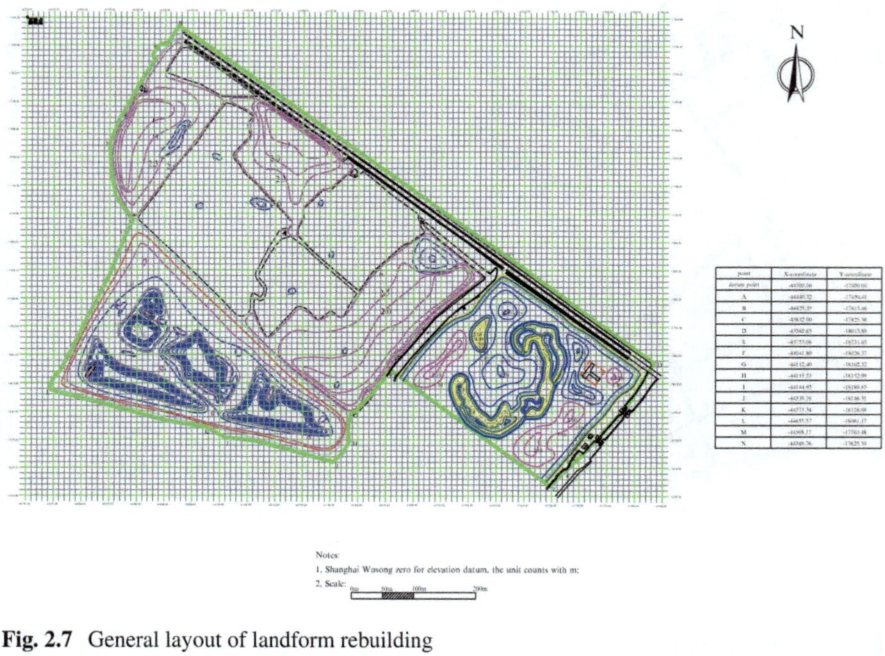

point	X-coordinate	Y-coordinate
datum point	-44700.00	-17400.00
A	-44440.32	-17450.41
B	-44425.35	-17815.46
C	-43632.00	-17825.36
D	-43565.65	-18013.80
E	-43753.06	-18231.45
F	-44041.80	-18026.37
G	-44112.40	-18102.32
H	-44115.53	-18352.99
I	-44144.95	-18180.65
J	-44539.79	-18186.35
K	-44373.34	-18128.08
L	-44655.57	-18061.17
M	-44508.17	-17563.48
N	-44549.76	-17625.30

Fig. 2.7 General layout of landform rebuilding

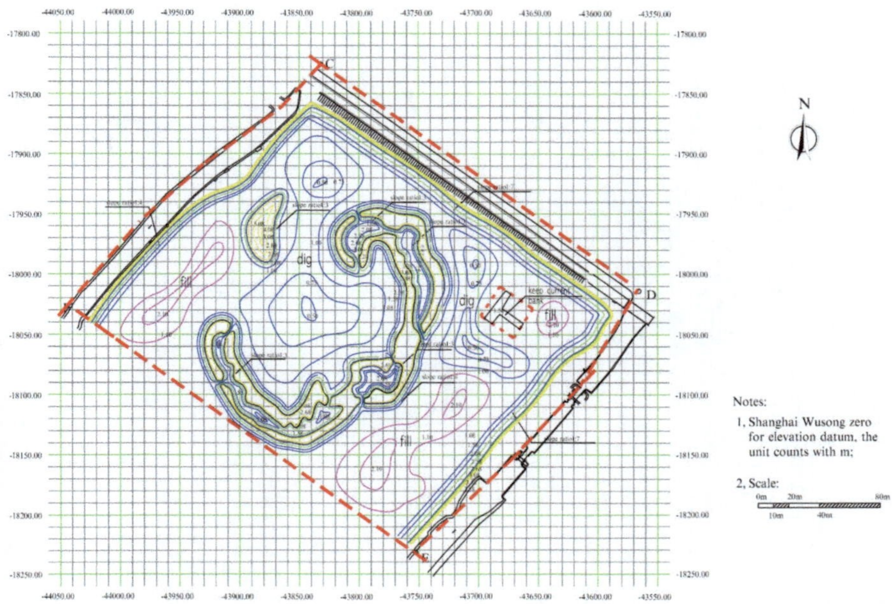

Fig. 2.8 Landform rebuilding plan of A area

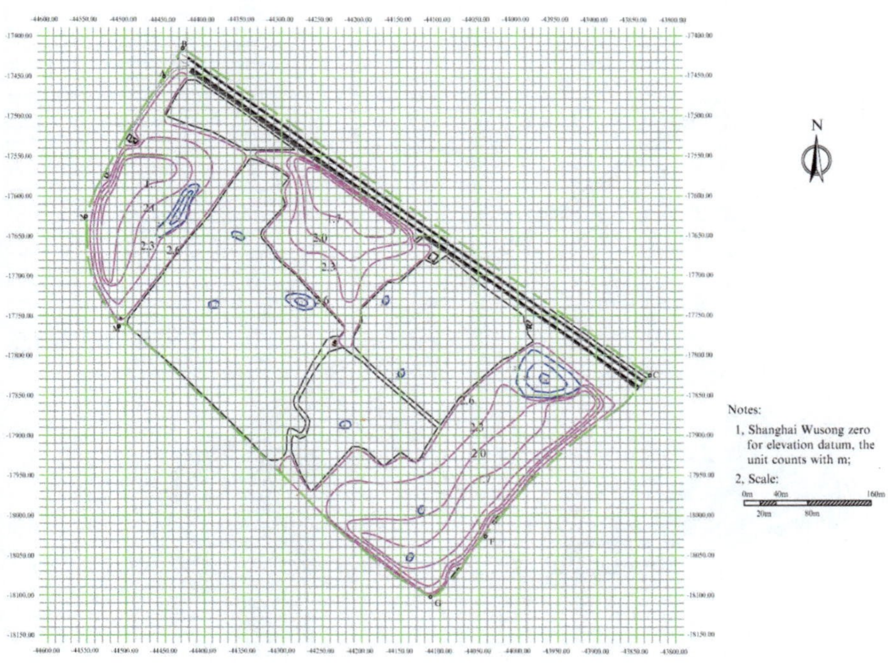

Fig. 2.9 Landform rebuilding plan of B area

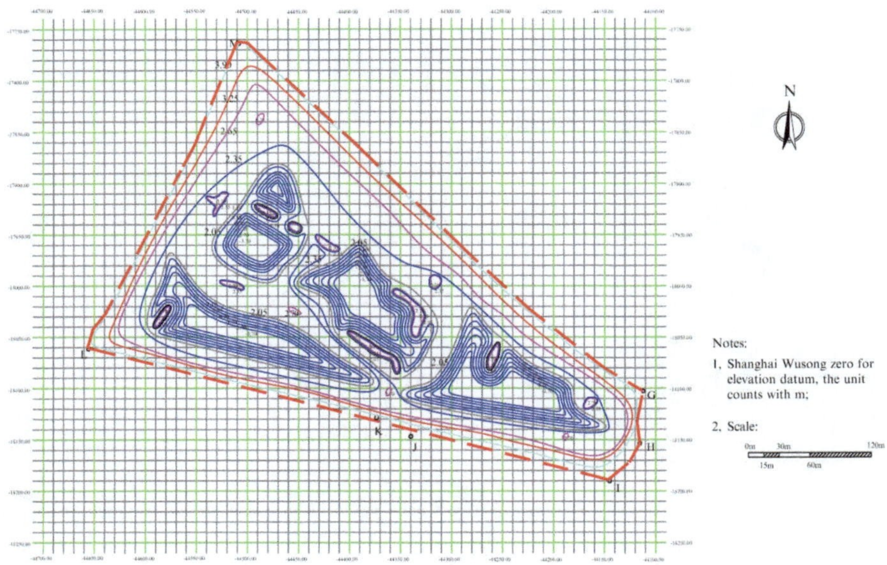

Fig. 2.10 Landform rebuilding plan of C area

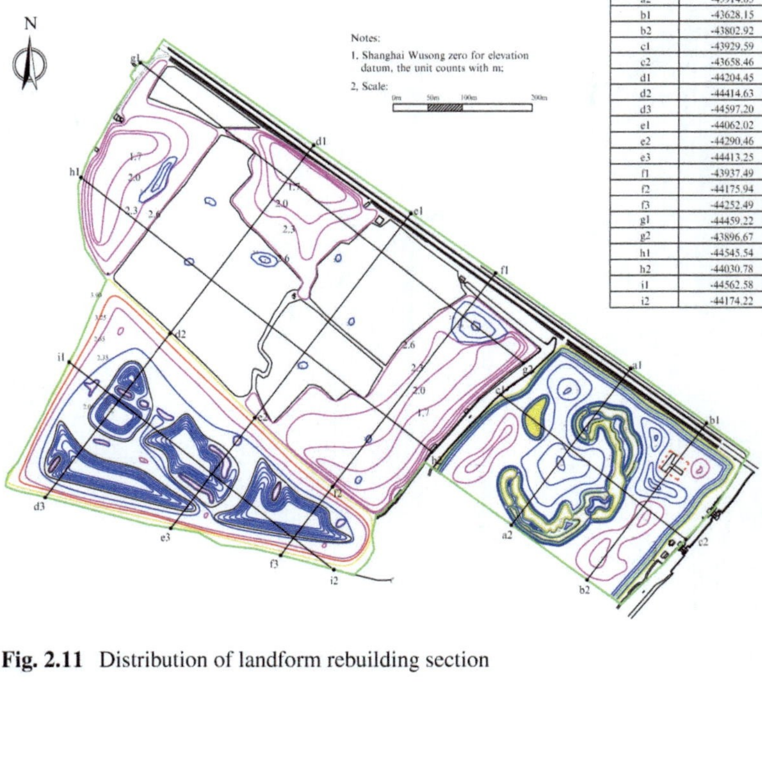

point	X-coordinate	Y-coordinate
a1	-43739.47	-17890.85
a2	-43914.05	-18117.20
b1	-43628.15	-17969.64
b2	-43802.92	-18196.22
c1	-43929.59	-17928.87
c2	-43658.46	-18137.99
d1	-44204.45	-17567.92
d2	-44414.63	-17840.43
d3	-44597.20	-18077.11
e1	-44062.02	-17666.37
e2	-44290.46	-17962.55
e3	-44413.25	-18121.74
f1	-43937.49	-17752.45
f2	-44175.94	-18061.59
f3	-44252.49	-18160.84
g1	-44459.22	-17449.02
g2	-43896.67	-17882.93
h1	-44545.54	-17614.74
h2	-44030.78	-18011.79
i1	-44562.58	-17882.00
i2	-44174.22	-18181.55

Fig. 2.11 Distribution of landform rebuilding section

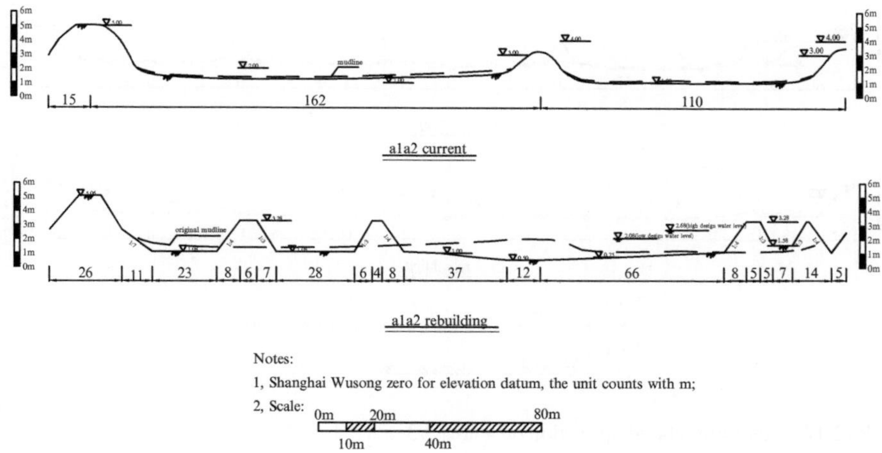

Fig. 2.12 Landform rebuilding section of A area 1/3

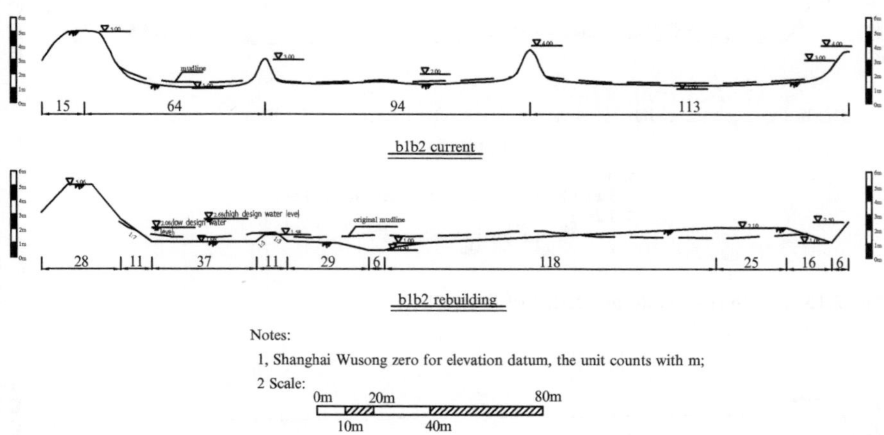

Fig. 2.13 Landform rebuilding section of A area 2/3

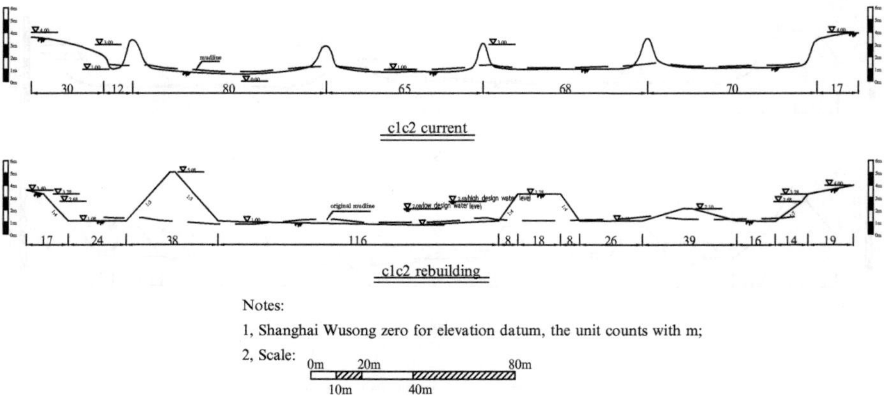

Fig. 2.14 Landform rebuilding section of A area 3/3

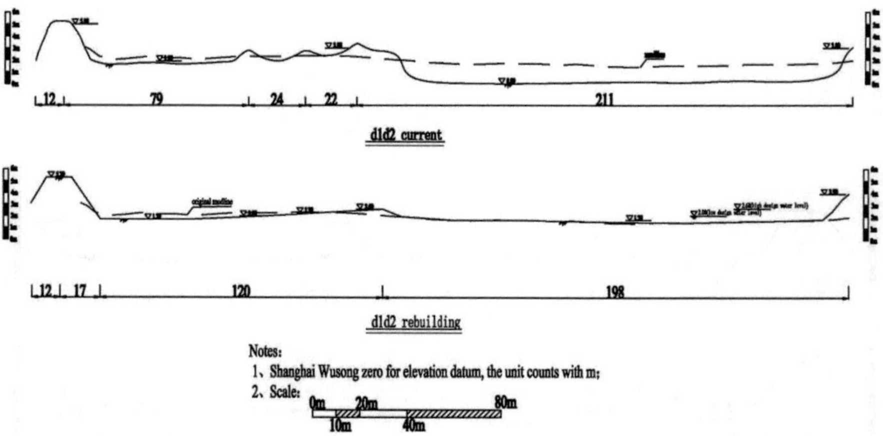

Fig. 2.15 Landform rebuilding section of B area 1/5

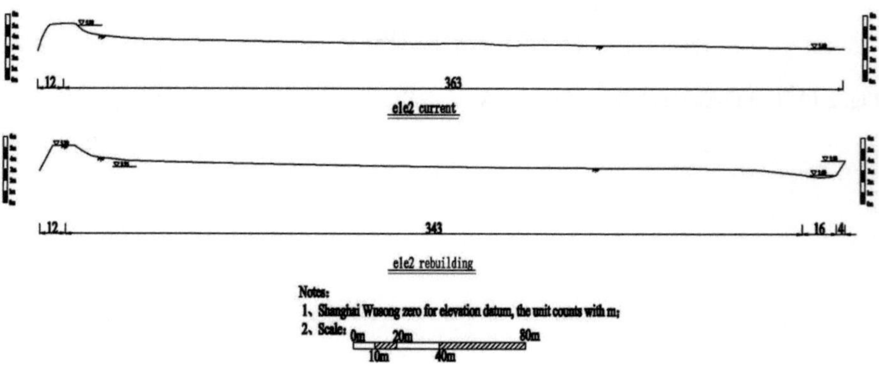

Fig. 2.16 Landform rebuilding section of B area 2/5

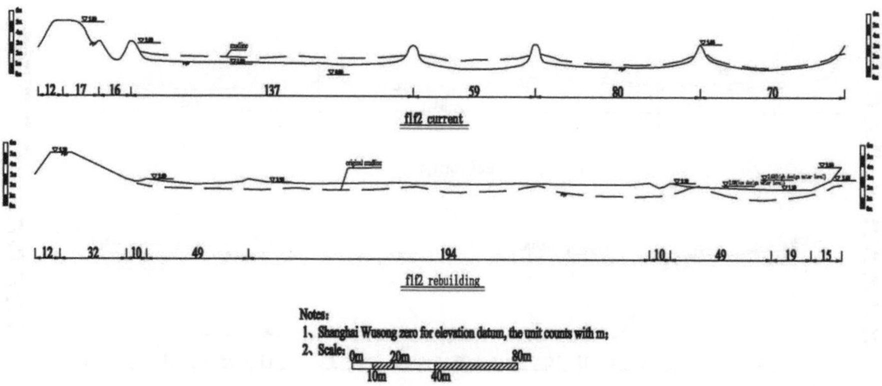

Fig. 2.17 Landform rebuilding section of B area 3/5

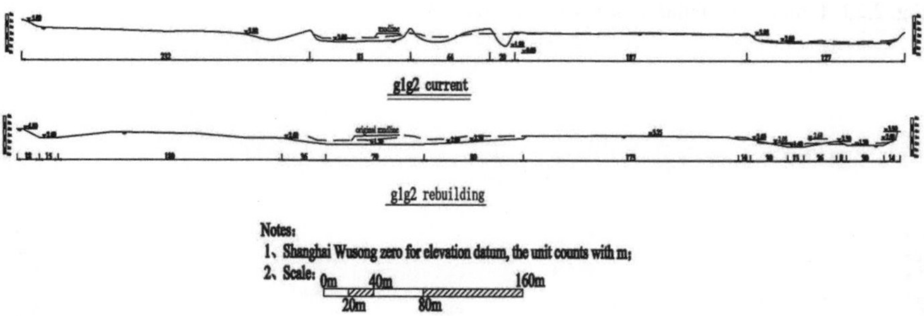

Fig. 2.18 Landform rebuilding section of B area 4/5

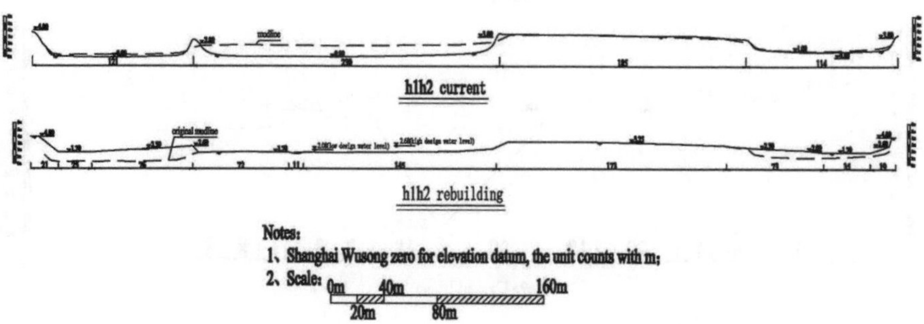

Fig. 2.19 Landform rebuilding section of B area 5/5

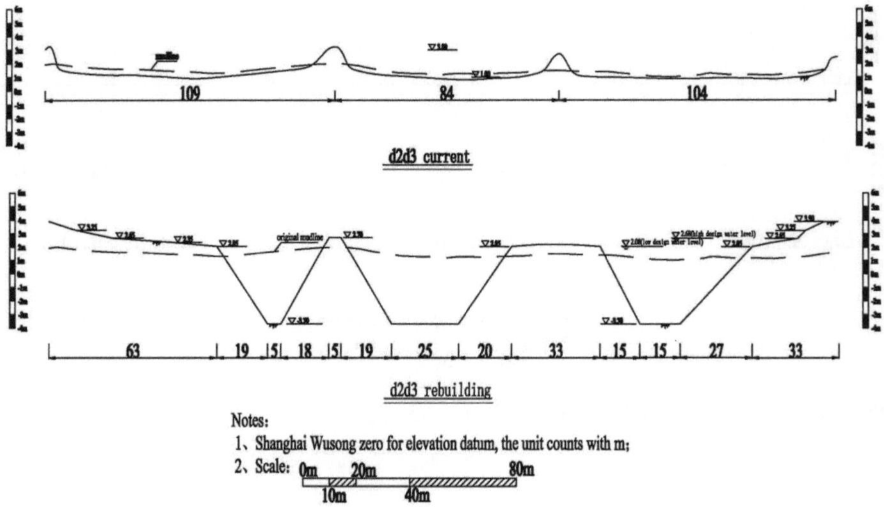

Fig. 2.20 Landform rebuilding section of C area 1/4

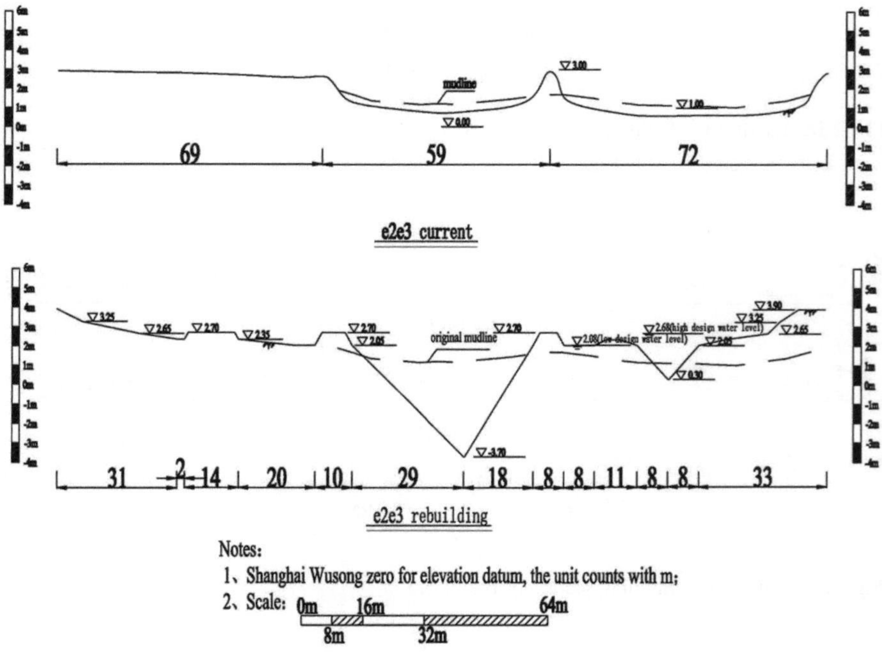

Fig. 2.21 Landform rebuilding section of C area 2/4

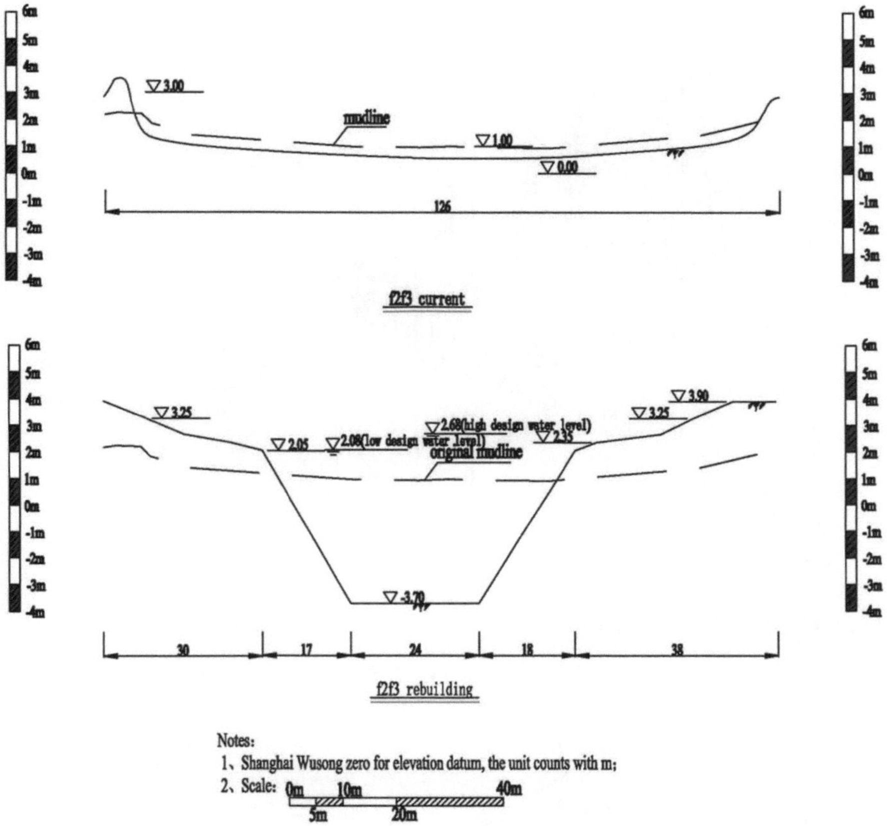

Fig. 2.22 Landform rebuilding section of C area 3/4

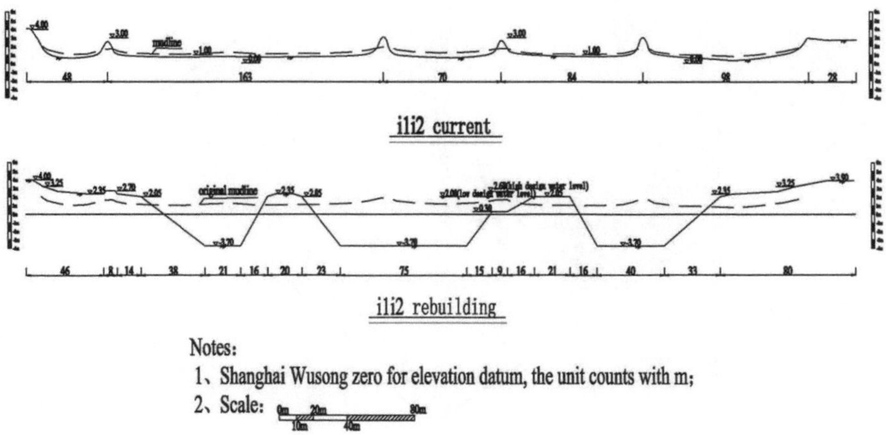

Fig. 2.23 Landform rebuilding section of C area 4/4

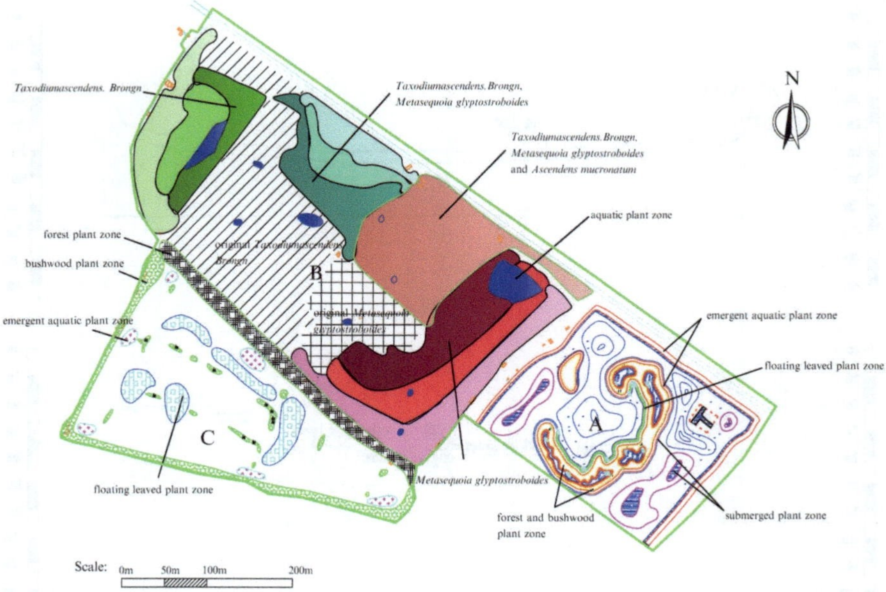

Fig. 2.24 General layout of plant zoning

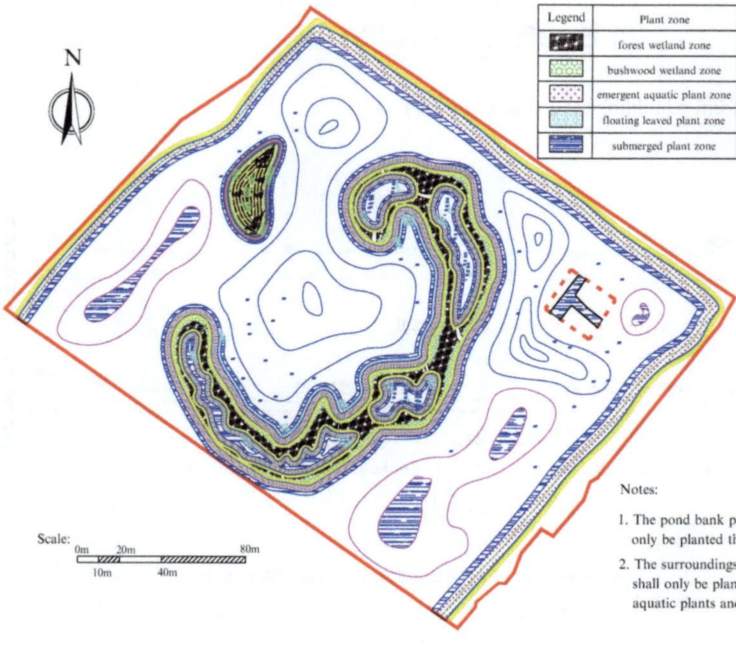

Legend	Plant zone	Area/m²	Density
	forest wetland zone	5220.6	1 plant/5m²
	bushwood wetland zone	766.0	2 plant/m²
	emergent aquatic plant zone	13903.3	7 plant/m²
	floating leaved plant zone	978.7	7 plant/m²
	submerged plant zone	9289.6	10 plant/m²

Notes:

1. The pond bank preservation zone shall only be planted the submerged plants
2. The surroundings of the project area shall only be planted the emergent aquatic plants and the submerged plants

Fig. 2.25 Plant zoning plan of A area

Legend	Plant	Status	Area/m²	Plantation gap/m
(hatched)	Taxodiumascendens.Brongn	original	55000	original
(cross-hatched)	Metasequoia glyptostroboides		17500	
d	Taxodiumascendens.Brongn Metasequoia glyptostroboides Ascendens mucronatum	new	33000	3~4
a1	Taxodiumascendens.Brongn	new	8400	2.5~4
a2	Taxodiumascendens.Brongn Metasequoia glyptostroboides		12000	
a3	Metasequoia glyptostroboides		18000	
b1	Taxodiumascendens.Brongn		8100	3.5~5
b2	Taxodiumascendens.Brongn Metasequoia glyptostroboides		6600	
b3	Metasequoia glyptostroboides		14000	
c1	Taxodiumascendens.Brongn		11800	4.5~7
c2	Taxodiumascendens.Brongn Metasequoia glyptostroboides		6700	
c3	Metasequoia glyptostroboides		18000	
(blue)	aquatic plant		5000	

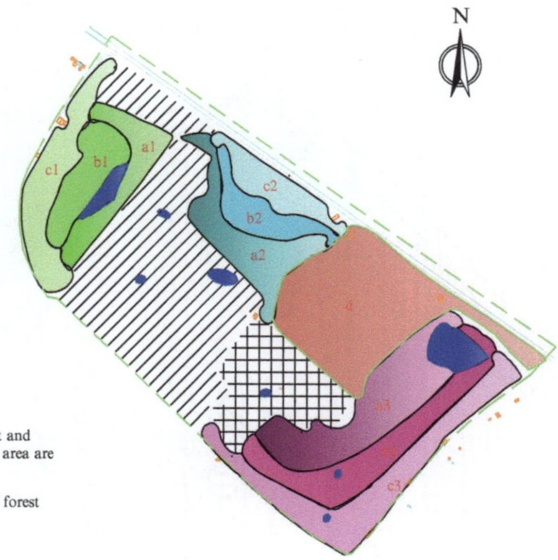

Notes:

1. The existing *Metasequoia glyptostroboides* forest and *Taxodiumascendens.Brongn* forest in the project area are preserved
2. The original *Cinnamomum camphora (L.) Presl* forest shall be replaced by *Ascendens mucronatum*
3. Scale: 0m 40m 160m 20m 80m

Fig. 2.26 Plant zoning plan of B area

Legend	Plant zone	Area/m²	Plant quantity
(forest symbol)	forest	11700	150
(bushwood symbol)	bushwood	10000	250
(emergent aquatic plant symbol)	emergent aquatic plant	3000	200
(floating leaved plant symbol)	floating leaved plant	10000	200

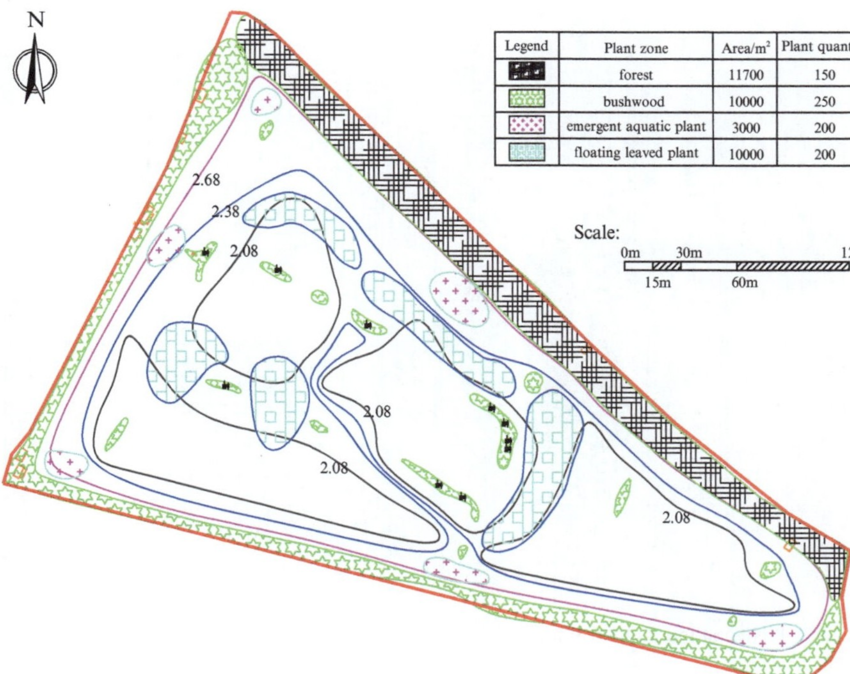

Scale:
0m 30m 120m
15m 60m

Fig. 2.27 Plant zoning plan of C area

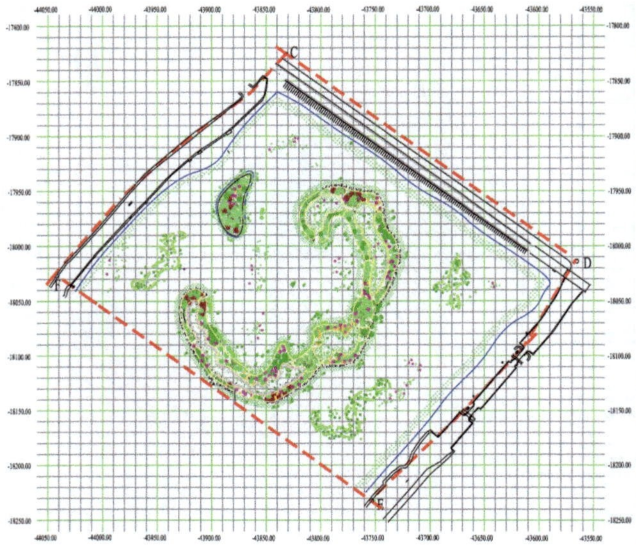

Legend	Species	Quantity	Size	Unit
	Taxodium distichum	50	D12	plant
	Taxodiumascendens.Brongn	100	D10	plant
	Metasequoia glyptostroboides	400	D12	plant
	Salix purpurea	100	D12	plant
	Pterocarya stenoptera C. DC	50	D10	plant
	Albizia julibrissin	50	D12	plant
	Sapium sebiferum	50	D12	plant
	Bambusa glaucescens	100	H100	plant
	Papermulberry	50	D12	plant
	Salix matsudana	50	D12	plant
	Glyptostrobus pensilis Melia azedarach Linn	100	D12	plant
	Morus alba L. Alnus trabeculosa	100	D12	plant
	Distylium buxifolium (Hance) Merr. Weeping Forsythia Rosa multiflora	300	H100	plant
	Winter jasmine Jasminum floridum Lonicera Japonica	300	H100	plant
	Osmanthus fragrans Forsythia viridissima Buddleja lindleyana	250	H100	plant
	Polygonum orientale Linn.	400	H100	plant
	Physalis alkekengi	100	H100	plant
	Pennisetum alopecuroides (L.) Spreng Setaira viridis (L.) Beauv	1500	Grass seed sows. grass and sod half-half	m^2
	Carex tristachya Schoenoplectus triguster (L.) Palla	1000	Grass seed sows. grass and sod half-half	m^2
	Cynodondactylon (Linn.) Pers Mat grass nutgrass flatsedge	1500	Grass seed sows. grass and sod half-half	m^2
	Typha orientalis Presl Acorus calamus Linn	1000	Young plant	plant
	Softstem bulrush Ludwigia peploides (Kunth) Kaven	1000	Young plant	plant
	Paragmites communis Trin	60000	Young plant	plant
	Arundo donax Cortaderia selloana	4000	Young plant	plant
	Wide Zizania caducifiora Oenanthe javanica (Blume) DC	10000	Young plant	plant
	Sagittaria sagittifolia L. Eleocharis dulcis	6000	Young plant	plant
	Alisma plantago-aquatica Lotus	3500	Young plant	plant
	Ipomoea aquatica Forsk Eomecon chionantha Hance	5500	Young plant	plant
	Hydrocharis dubia (Bl.) Backer Pumilura	2000	Young plant	plant
	Fructus Trapae Quadrispinosae Euryale ferox Salisb. ex DC	2000	Young plant	plant
	Potamogeton distinctus A. Bennett Potamogeton crispus Vallisneria natans (Lour.) Hara	50000	Young plant	Bushwood
	Myriophyllum spicatum Hydrilla verticillata	40000	Young plant	Bushwood

Fig. 2.28 Plant configuration of A area

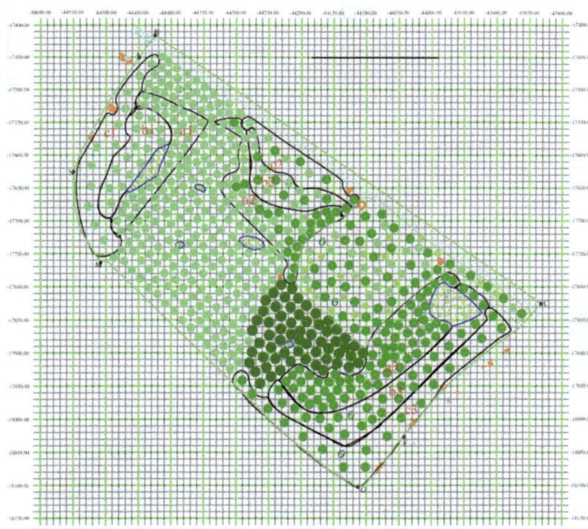

Legend	Species	Quantity	Size	Unit
	Taxodiumascendens.Brongn	5000	D10	plant
	Metasequoia glyptostroboides	8500	D10	plant
	Ascendens mucronatum	6000	D10	plant
	Euryale ferox Salisb. ex DC	5000	young plant	plant
	Nymphaea alba	5000	young plant	plant

Fig. 2.29 Plant configuration of B area

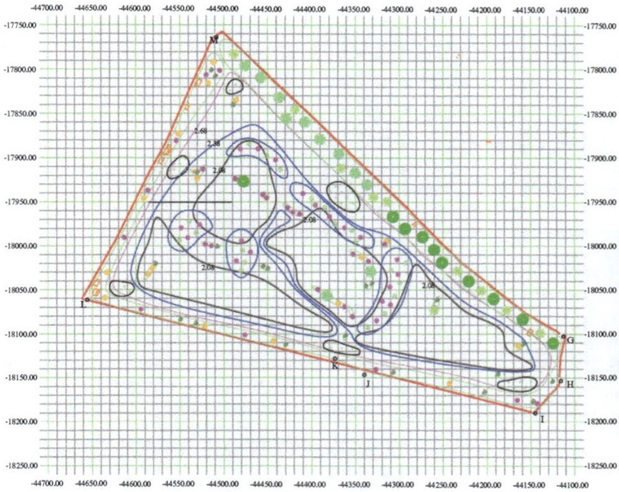

Legend	Species	Quantity	Size	Unit
	Taxodium ascendens Brongn	50	D12	plant
	Metasequoia glyptostroboides	50	D12	plant
	Ascendens mucronatum	50	D12	plant
	winter jasmine Lonicera Japonica	100	H100	plant
	Osmanthus fragrans Forsythia viridissima	100	H100	plant
	Polygonum orientale Linn.	50	H100	plant
	Phragmites communis Trin	200	young plant	plant
	Hydrocharis dubia (Bl.) Backer Pumilum	100	young plant	plant
	Furctus Trapae Quadricaudatae Euryale ferox Salisb. ex DC	100	young plant	plant

Fig. 2.30 Plant configuration of C area

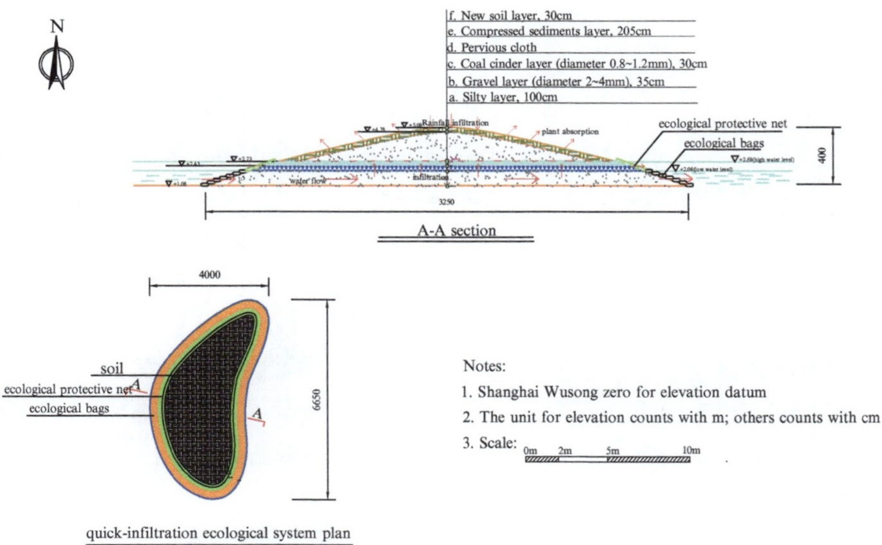

Fig. 2.31 General layout of quick-infiltration ecological system

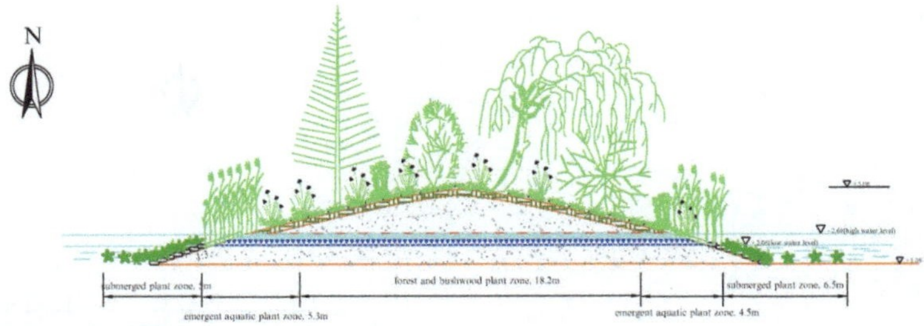

A-A plant section

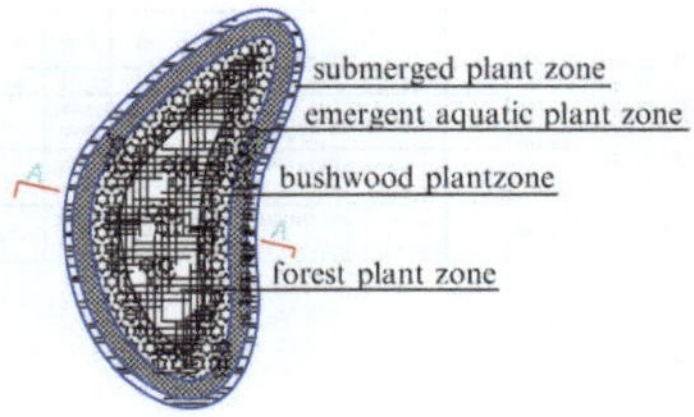

quick-infiltration ecological system plan

Notes:

1. Shanghai Wusong zero for elevation datum
2. The unit for elevation counts with m; others counts with cm
3. Scale: 0m 2m 5m 10m

Legend	Plant zone	Area/m²
	forest plant zone	900
	bushwood plantzone	100
	emergent aquatic plant zone	500
	submerged plant zone	250

Fig. 2.32 Plant configuration of quick-infiltration ecological system

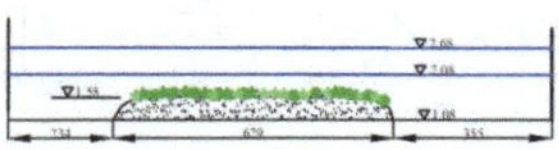

submerged plant zone

Notes:

1. Shanghai Wusong zero for elevation datum
2. The unit for elevation counts with m; others counts with cm
3. Scale: 0m 1m 2m 4m

Legend		
	Pterocarya stenoptera C. DC	forest wetland zone
	Salix matsudana	
	Metasequoia glyptostroboides	
	Rosa multiflora	bushwood wetland zone
	winter jasmine	
	Forsythia viridissima	
	Phragmites communis Trin	emergent aquatic plant zone
	Cortaderia selloana	
	Wide Zizania caduciflora	
	pumilum	floating leaved plant zone
	Furctus Trapae Quadricaudatae	
	Euryale ferox Salisb. ex DC	
	Hydrilla verticillata	submerged plant zone
	Myriophyllum spicatum	
	Potamogeton crispus	

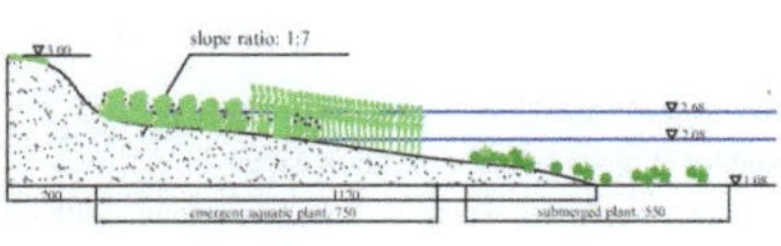

ecological revetment 1

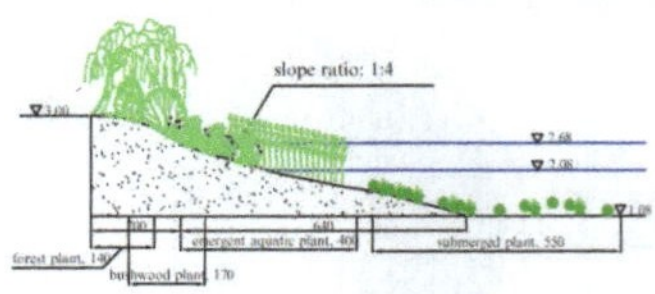

ecological revetment 3

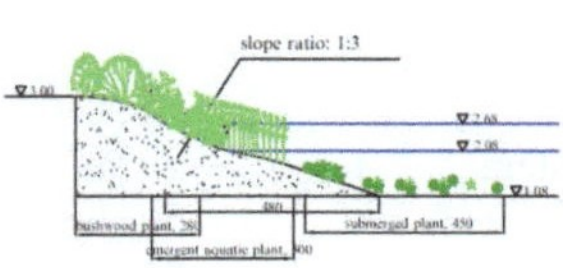

ecological revetment 2

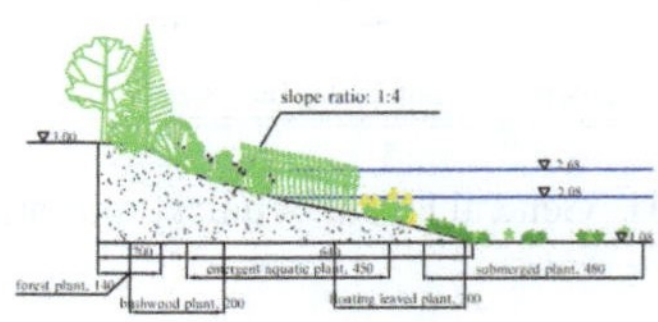

ecological revetment 4

Fig. 2.33 General layout and plant configuration of revetment in A area

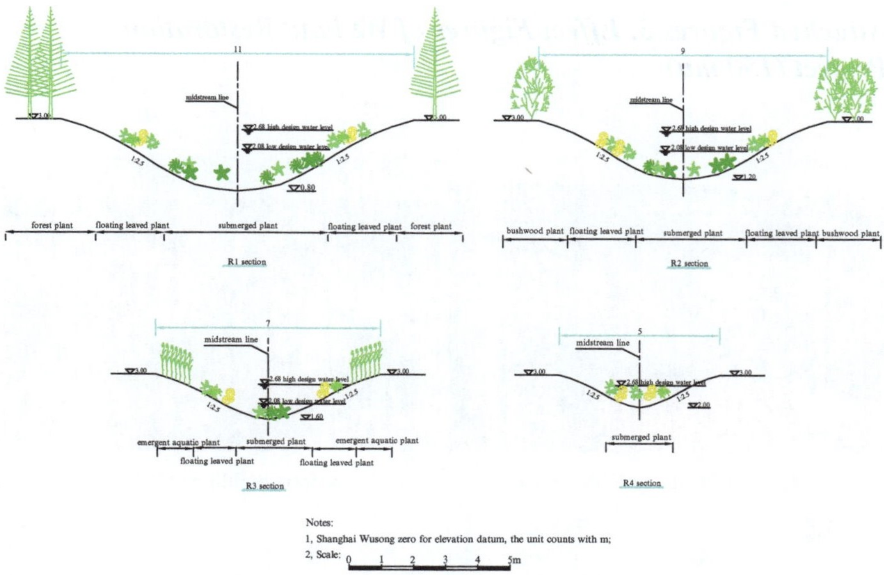

Fig. 2.34 General layout and plant configuration of revetment in B area

Attached Figures 3. Effect Figures of Wetland Restoration Project (150 mu)

1. Current situation (2008. 08. 01)

2. Landform building (2009. 03. 20)

3. Completion of landform building (2009. 04. 22)

4. Vegetation restoration in the large island (2009. 04. 28)

5. Completion of shoal (2009. 05. 06)

6. Aquatic vegetation restoration (2009. 05. 29)

7. Beginning of the quick-infiltration ecological system (2009. 06. 15)

8. Construction of the quick-infiltration ecological system (2009. 06. 26)

9. Completion of the quick-infiltration ecological system (2009. 07. 31)

10. Vegetation restoration in small island (2010. 02. 25)

11. Completion of project (2010. 03. 29)

12. Effect of wetland project (2010. 06. 02)

13. Effect of wetland project (2010. 06. 05)

14. Effect of wetland restoration project (2010. 06. 05)

Postscript & Testimonials

The book titled by "Wetland Restoration: Shanghai Dalian Lake Project" will be published, is it a success, delighted or anxious? I felt I am facing with a thousand of complicated thoughts and all kinds of feelings, so I am trying to write done something for memory…

Rebuilding the Water Source of Life

Standing by the Dianshan Lake, a new picture of the south of the lower reaches of the Yangtze River emerges before you; which makes you excite and go beyond your dream as the important Huangpu River water source protection area, 625 mu fishponds have been restored, a small and a large ecological island is rightly located opposite each other. By looking down in air, it looks like a shoe-shaped gold ingot. The two islands are almost 3,000 m high; the local water is over 2 m deep; where the water is purified by the different vegetation zones from top to bottom, and various aquatic animals are bred in the water; all of these plants and animals form a perfect wetland biologic chain. The tall and straight pond cypress trees are not alone any more, the luxuriant trees in the wetland makes the pleasant shade everywhere. In the forest, there are winding streams, colorful ponds, water birds, wild ducks, egrets, even migrant birds are attracted to come here for passing the winter every year. The local people call Dianshan Lake as "land of idyllic beauty". Meanwhile, the "water forest" has the characteristics of preserving water source and absorbing carbon sink; therefore it is very worthy of being protected. By maintaining the water of Dalian Lake run always, its circulation and convergence through the water intake and Huangpu River can hugely improve the drinking water quality of Shanghai area. In Cenbu Village, the local peasants do not drain domestic sewage into the rivers and discard rubbishes directly. In 2009, the village was recognized as the nationwide demonstration village building a new socialist countryside of China; the biggest highlight of the village is the one that all agricultural and leisure activities are conducted in the natural way, such as the natural agriculture,

S. An and L. Wang, *Wetland Restoration: Shanghai Dalian Lake Project*,
Springer Environmental Science and Engineering, DOI 10.1007/978-3-642-54230-5,

agricultural production without using fertilizer or pesticide, and organic agriculture product production. The mode to let townsfolk adopt the vegetable field, conduct the leisure participation and launch the dispatching services can not only support the local development, but also can efficiently constrain the passive situation that the water source in the agricultural non-point source is polluted.

This is the Dalian Lake life water source pattern created by Shanghai.

Heavy Responsibilities

The Dalian Lake Mode is an integrated water source area governance pattern based on the agricultural non-point source pollution in Tai Lake basin. Featuring in multiple characteristics, such as leading ecological rehabilitation technology, organic agricultural production, long-acting control on community environment, ecological compensation policy innovation, the pattern breaks through the passive situation of the water source area protection and development. In accordance with the third-party's independent monitoring report of the professional department, the water quality in the demonstration area keeps on complying with the nationwide drinking water quality standard (Grade II). In addition to the rich biodiversity and 40 species of animals and plants in the local wetland, over 10 migrant birds are attracted for inhabitation. Different from the previous governmental governance engineering, the community participation advocated by WWF is actively responded by all stakeholders; the local peasants adopt the ecological agriculture concept and technology, so the local farmlands are not applied fertilizers and pesticides, which makes the local farmlands and fishponds realize the zero emission. Meanwhile, by taking the opportunity that Shanghai conducts the sewage and waste pipe construction, as well as the ecological & economic compensation policy of Shanghai Municipality has been gradually implemented and WWF launches and boosts the "Water Source Partner 1+1" program, the residents of local community have outstanding improved their environmental awareness and enthusiasm, and the long-acting water environment guarantee mechanism has basically formed and keep on improving.

However, all of us should be aware of the unprecedented impact suffered by the countryside in the current rapid urbanization process, *i.e.*: the loss of young and postadolescent laborers, dissolution of traditional community mechanism, weakened neighboring relationship; in sense of the development orientation, the role of the countryside becomes fuzzy day after day, even the countryside becomes the land use indicator of urban and industrial growth. Meanwhile, the huge change of agriculture and countryside impacts the sustainable development of urban and township economy; among which the biggest impact is the food safety. The misuse of pesticides, fertilizers and food additives, as well as the extensive agricultural production mode has already threatened the food safety. As an integral part of the complicated socials system engineering of water source area protection, Dalian Lake pattern just starts its trip. Seeking the path of economic development, traditional culture's protection and inheritance, highly harmony of human and nature, rebuilding the beautifully scenes of the south of the lower reaches of the Yangtze River, rehabilitating the damaged countryside and social cells can not be done in a day; in future, there is a long road before the popularization of the pattern in the upper reaches of the Huangpu River and the whole Tai Lake Basin.

MIX
Papier aus verantwortungsvollen Quellen
Paper from responsible sources
FSC® C105338
FSC
www.fsc.org

If you have any concerns about our products,
you can contact us on
ProductSafety@springernature.com

In case Publisher is established outside the EU,
the EU authorized representative is:
Springer Nature Customer Service Center GmbH
Europaplatz 3, 69115 Heidelberg, Germany

Printed by Libri Plureos GmbH
in Hamburg, Germany